The IMAE Business & Academic ERP Implementation Series

HOW TO CONSIDER, SELECT, AND IMPLEMENT AN ERP SYSTEM

GUIDE ONE

Dr. Jill O'Sullivan • John Rico • Dennis Goldensohn

IMAE Publications
A Division of DMMSI Associates, Inc.

IMAE Business & Academic ERP Implementation Series

How to Consider, Select and Implement an ERP System

by Dr. Jill O'Sullivan, John Rico and Dennis Goldensohn

For Product Information, or Permissions as expressed above, Contact DMMSI at:
DMMSI Associates, Inc., 320 Carleton Avenue, Suite 6200, Central Islip, NY 11722
www.instituteforimae.com

ISBN: 978-0-9912142-2-8
LOC Control Number: 2015900763

Screenshots: courtesy of Synergy Resources LLC, MS Office Suite 2010, Infor Visual 7.1.2.
Permissions granted by INFOR® | Editor/Interior Layout/Cover design by Bernita J. McGoldrick

FOREWORD

Whether you have an ERP system or are looking to replace your legacy system with a modern ERP system solution there are many benefits associated with this consideration. Some of these are improved customer service, cash flow, operations management, quality, production and visibility. Continuous improvement will be an easier task with the proper use of the ERP system.

Undertaking an ERP project requires that there is going to be total commitment from upper management, and a very well structured project plan. In considering the needs the company has for this ERP solution, the company must understand what the differences are between the many ERP solutions providers, and what the goal and deliverables will be.

This ERP Guide will break this down into simple and manageable steps so that you can take the project from its inception through the selection of an ERP software provider-reseller to the implementation process. This will give your company the ability to use this new investment for better customer service, improved process controls, increase in customer retention, reduced lead times, improved inventory management, which will also enhance cash flow. Although this may seem like a daunting task, if the project is broken down into manageable steps this will significantly help the end user select, prepare, and implement an ERP solution for your company.

ICON GUIDE

This symbol states that this is a very important concept that should be considered and should be followed!

This symbol tells that awareness should be made of 'what can happen' and 'why.' Take this into consideration when moving on in the implementation process.

This symbol recognizes an area of hazards in ERP thinking.

Since this is important, the concept stated should be reviewed to understand why it is used in the implementation process and what the outcomes might be.

Before going any further—STOP. Know that this symbol stresses this concept in this section is important and why!

This symbol represents an idea that should be considered in the implementation process.

This symbol represents documenting procedures, policies or work instructions.

This symbol represents an opportunity to think about how this information can help you.

***Important parts of this guide that will be highlighted in NOTES Boxes.**

ABOUT THE AUTHORS

Dr. Jill O'Sullivan is an Assistant Professor at Farmingdale State College in New York, where she teaches ERP, Management Information Systems, Operations and Supply Chain Management, and Strategic Management. She is first past president of APICS NYC/LI and the Faculty/Student Advisor of the Student Chapter at Farmingdale State College. She serves on the Advisory Editorial Board of the Journal of Systemic, Cybernetics, and Informatics (JSCI). Dr. O'Sullivan serves on the UASGC CTE Advisory Board, the Regional Industry Council on Long Island P-Tech Industry Committee, and is an original member of the Manufacturing Executive Leadership Council, servings as Judge for the ML Awards. Jill is the Founder/President of JJKT Consulting Services and previously founded JJK Sales, a Manufacturers Sales company representing the northeastern chapter regions. Her company specialized in custom manufacturing, assembly and electronic component products for a diverse customer base of Industrial/OEM, Distributor/Wholesalers, Value Added Reseller, Military/Aerospace, Government and Medical companies. Formerly a Purchasing and Materials Manager on Long Island, Mrs. O'Sullivan has acquired more than 24 years of experience in Operations and IT in military, aerospace, government and commercial companies. Jill is married to John and has two sons, Kyle and Tom. (Oliver)

John Rico has been in the Manufacturing, Production and Materials Management field for over 30 years. He has worked in several different types of business environments that include: Aerospace, Pharmaceutical and consumer based products. John has been through several major ERP implementations, of which he chaired two very successful projects and served as a task team leader on another. John has been through many software upgrade implementations and ISO 9000 implementations. He has seen successes and failures and as a result knows what works and what does not. John received my certification from American Production and Inventory Control society as CPIM early in his career and it has served as a solid foundation for all his future learning experiences. John has been through a great deal of formalized training in preparation for implementations and "Train the Trainer" programs.

Dennis Goldensohn has been employed in the manufacturing industry for the past 44 years, with the past forty one years with Arkwin Industries, where he serves as the Production Control Coordinator for the Manufacturing facility. Dennis lends his expertise to Arkwin in the areas of Lean facilitator, trainer, and the company photographer. Dennis was also part of Arkwin's ERP Implementation team. Dennis is a graduate of Farmingdale State College with a BA Degree in the Management of Technology, Sigma Beta Delta, and recipient of the Farmingdale-SUNY Alumnus of the Year award for 2012 for his contributions to institution of Education and continuing professional development. Dennis is a member of the APICS New York City-Long Island chapter and is the past president of the chapter. He is a sought after guest speaker in the areas of manufacturing, particularly in the field of Lean training. Dennis is a member of the Long Island Society of Manufacturing Engineers-SME chapter. Dennis serves in the P-Tech program as the chairperson of the Industry Partner subcommittee. Dennis and his wife, Ellen, have a son and a daughter, and two grandchildren, where they have lived in Levittown, New York, for the past 28 years. Dennis is an avid semiprofessional photographer. Dennis has been a dedicated volunteer for the Diabetes Research Institute for the past 30 years.

TABLE OF CONTENTS

CHAPTER 1
ERP/MRP/MPS

OVERVIEW: SYSTEM SETUP (SYSTEM VARIABLES OR PARAMETERS)

Before we get into some detail on important inputs to your ERP system it is vital we speak a little bit concerning setup. All systems have, what is known as, Parameters or System Variables that need to be reviewed and set based on your business model. There can be as many as 60 to 100 or more of these system variables and it is critical to understand their influence on ERP output. Some will be as simple as setting up your calendar and defining whether or not you will be functioning in calendar days or business days. Others, more complicated, will ask if you want the system to generate planned orders when a demand exists or to first recommend a reschedule of an existing order if one exists. You will need to speak to the software provider and get a list of these variables, understand their impact to system output and decide how best to set them. This needs to be done with the input of the entire steering committee. (Chapter 2) The variables are often process related and will pertain to one or more aspects of the overall ERP process.

The entire ERP implementation budget requirements must be considered for evaluation and approval.

Enterprise Resource Planning or ERP is the most recent evolution of the original Material Requirements Planning or MRP. Material Requirement Planning sought to fulfill a fundamental need and that is to become more effective and efficient in managing material flow. At its core, it is meant to manage the time phased receipt of material based on actual or forecasted demand. In its simplest form, the time phasing of material receipts is determined by the specific build requirements for the end item. This is the sequential chain of events that is required to take a group of individual parts and put them together to create the item that is sold to the customer. This time phasing is a function of the end items Product structure.

The product structure represents a bottom up build from individual part numbers through sub assembly build to final assembly. Final assemblies and sub-assemblies are made up of either purchased parts or make items that are assembled or manufactured in house or a combination of the two.

The respective lead times associated with these builds is what creates the linkages between sub-assemblies and their parent assemblies so that a lower level sub-assembly's due date must be the start of the parent assembly it goes into. These lead times are often determined by engineering time studies to determine an efficient and reasonable amount of time it should take to build an assembly. Once established, these time studies result in a "Standard Time" that is the goal for the assembler to meet or exceed.

Lead time for the purchased material is often determined and set by the Purchasing department based on current market conditions, as defined by the supplier/s of the purchased part. The first time buy for these purchased parts are often used to create a "Standard Cost" which all future purchases will be judged against. Any variance to this cost is known as the Purchased Price Variance or PPV. The due date of these parts is determined by the start date of the assembly it is required on.

The date that the part must be placed on order is backed off from the start date of the assembly it is required on by it specific lead time. The lead time, internal or external, is often stated in number of days required to place the part on order and get an acceptable part in house and delivered to the assembly it is required on. Or reflective of the time it takes to start an assembly build and complete it. Cumulative lead time is a function of the longest lead time part, plus the lead time of the assemblies it goes into, depicted in the below diagram. The scheduling of the end item and its associate product structure is what determines the need date of the individual items in the completed assembly that is now ready for shipment.

Figure 1.1 Cumulative Lead Time

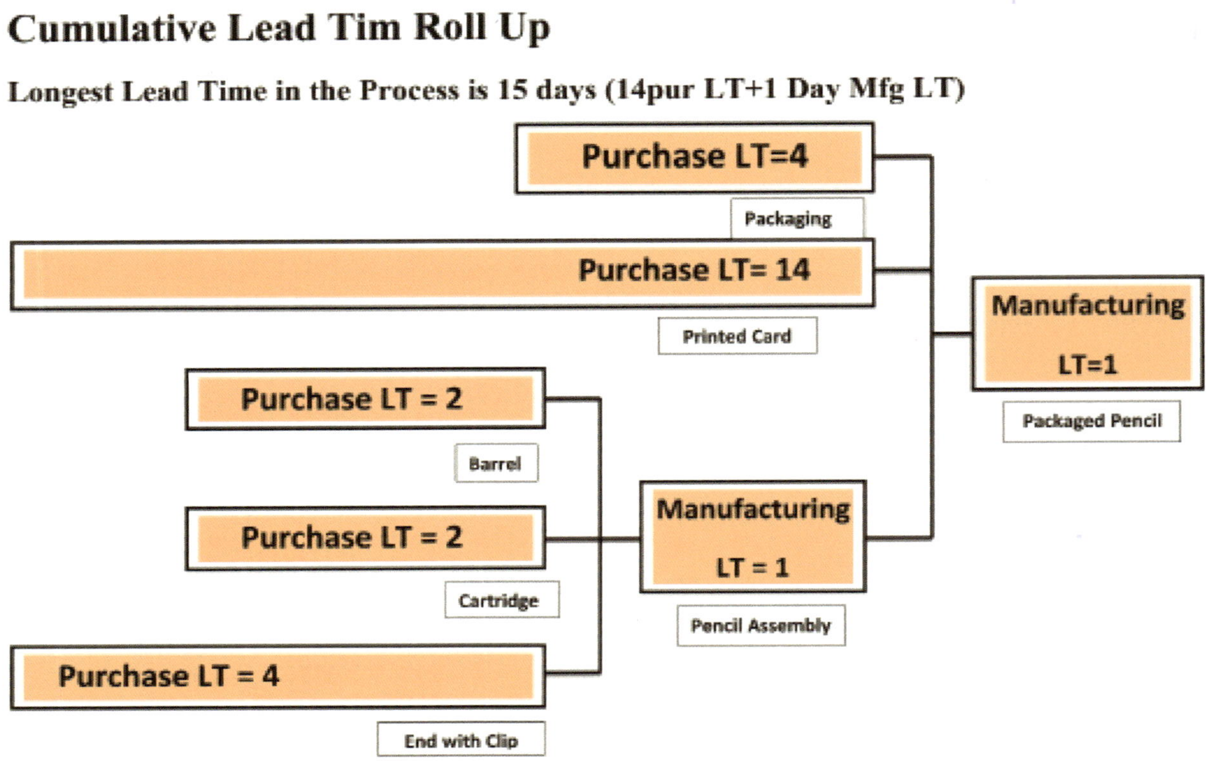

MRP II

The next evolution of MRP was MRPII, in MRPII the MRP stands for Manufacturing Resource Planning. This added another level of complexity by incorporating Capacity Planning of your internal resources such a machining equipment like the Lathe, CNC, Injection Mold equipment, shop floor labor also known as direct labor due to their direct value added contact with the product assembly process etc. Value added simply means to change a product in a manner that makes it more valuable to the end user or customer.

ERP

Enterprise Resource Planning or ERP is the culmination of many years of these concepts evolving into a full blown treatment of an end to end company-wide management software tool. An ERP system is meant to give you the tools, based on best known business practices, to manage your business from marketing to acquire sales for your product to management of after sales service and everything in between. There are many critical milestones along the way to a successful ERP environment and creating accurate and reliable data output.

There are what we call "Non-Negotiable" requirements that must be met that provide the foundation you need to build your ERP system upon. It is best to make sure you are doing the fundamentals correctly and accurately before undertaking an ERP implementation. A Strong foundation will ensure a successful ERP implementation. These key areas must be identified, procedures put in place to ensure consistency and the results must be quantifiable, measurable and repeatable.

The importance of accurate and timely data cannot be over-emphasized. There is an old computer phrase that states Garbage in Garbage out (GIGO). This phrase rings forever true and should especially be taken to heart when migrating to an ERP software system. If your existing data is inaccurate and unreliable a new software system will not fix that. Only improved systems, policies and procedures along with the proper training and education will do that.

The following are some fundamental inputs into an ERP system that must be accurate going in, to provide the strong foundation needed to be successful. Depending on your specific business model you may add to this list.

Inventory Accuracy: To begin with, inventory accuracy would be the most obvious one, but we are often surprised at how little attention is given to this very crucial aspect of creating a solid foundation for your ERP/MRP/MPS system. A proper cycle counting procedure goes a long way toward maintaining a high level of accuracy in your inventory and by high level we are looking at 98% or better. We will go into much more detail in a later chapter on this subject.

Bill of Material Accuracy: The allocation process is driven in great part by the bills of material used to identify your product build requirements inclusive of in house make items and externally purchased items. Bill of Material accuracy must be at 99% and audited to maintain that level. We will also drill down deeper into the methods used to maintain this high level of accuracy.

Work Order Management: The management of work orders from "Cradle to Grave" as it is often referred to, is also critical in maintaining system accuracy and reliability. This is the movement of material into and out of a work order that is often affected by non-conforming material that may have to be rejected, scrapped for whatever reason or shrinkage, which will often be built into the usage requirement of a bill of material for a work order. Labor reporting and number of pieces reported as complete must be accurate and audited.

Manufacturing Routings: Routings define the path material takes while being processed into a usable item for the next higher assembly or as it is made into the end item for sale to a customer. Routing accuracy is very important (99%) as it defines where parts should be during its build cycle.

The following website has a good presentation of ERP flow:
http://www.mcsohio.com/pdf/implementing_erp_sys.pdf

The longer the build cycle the more critical this becomes, as it pertains to time phased receipt of material and at what sequence in the multitude of routing operations a part may be required. There may be months between operations, as with the building of a Jumbo Jet. Routings also contain the much needed information required for "Capacity Planning." There will be both labor hour content and machine hour content. Engineering (manufacturing engineering) determines the standard labor or machine times required, based on time studies, and this represents your labor or machine load vs. your stated capacity. Details on how to capture this information and create a process for ensuring a high level of accuracy will also be treated in depth in a later chapter.

Change Management: We all know nothing ever changes in a manufacturing environment Right! Yeah right. The only certainty in Manufacturing, is change. If you do not have an effective way of managing change you will never be cost effective in the implementation of required changes. Poor change management can cause you to scrap too many parts or build too much for inventory or worse, expose you to product liability and lawsuits. All these mismanagement issues have dollar signs associated with them. Engineering Change Notices or ECN's are an effective tool for managing change when used properly.

Sales Forecast Accuracy: Anybody with an inadequate amount of knowledge with tell you that forecast by their very nature will be inaccurate. What is important is that it is measured, as a Key Performance Indicator (KPI). The accuracy of anything begins to improve the moment you begin to measure, track and report on it. Someone needs to be responsible for managing and reporting on the accuracy of the forecast and it should be done in several different ways depending on your business model. You may want to report on accuracy by sales person, product line or sales region. It is recommend that at least two or three different ways of tracking and reporting be required, on this all important input to your Master Production Schedule (MPS), if you have one, or your MRP work order system. We suggest using Booking versus Forecast to arrive at this number and we will elaborate on this in other chapters.

Master Production Schedule (MPS): The MPS should be closely aligned with the Sales Forecast. However, make no mistake it is the liaison between the Sales Forecast and your production work orders. Its function is to translate the Sales Forecast into a viable production schedule that supports the customer requirements, while taking into account shop floor constraints. The term Linearity comes into play here and we use this to try and level load the MPS in a fashion that the company constraints can manage. However, make no mistake, the MPS must support the Sales Forecast and customer demand. You NEVER, EVER schedule your work load to accommodate your constraints at the expense of satisfying your customer or for any reason for that matter. A work order for a master scheduled item should never be rescheduled unless the customer demand that it was created for moves in or out, then it should follow in kind, this is commonly referred to as "Back Scheduling." When you keep moving your MPS to accommodate your constraints you mask any issues, which may need to be addressed, this is what is referred to as "Chasing Supply" this is often also referred to as Forward Scheduling or Finite Scheduling. You never chase supply, you chase demand and you manage supply. Supply can mean, purchased parts, and externally produced parts, internally made items, internal machine or labor constraints. Once you start to chase supply you mask your problems and lose any sense of priority within the overall system. Once you mask your problems you cannot quantify them and if you can't quantify them you can't solve them. This method of scheduling to your constraints also removes management's ability to intervene, it also eliminates your company's ability to project grow and prepare for an expanding operation or even to identify it. Scheduling to your demand gives you the ability to plan future work and labor load to support growth. It also can demonstrate and help cost justify the need for more or better equipment and labor. It will help you to identify short term bottlenecks and long term growth opportunities.

Product Structure: Product structure, as it relates to MRP, is the configuration of a product and depicts the bottom up build of a product that is produced for final sale to the customer. The "For Sale" product or the end item to the customer is at the top of the pictorial depicting the build sequence from the bottom up consisting of both Make and Buy items. Depending on the company and their internal resources e.g., do they have access to an internal machine shop, sheet metal shop or a welding department, the decision to make an item or buy is mostly based on cost effectiveness. If you have the ability to make an item you must cost it out against the external cost to buy the item. Other factors such as quality and lead time also play a significant role in this decision process. The make or buy decision should be made during the design phase of a product. The final assembly item is often referred to as the Zero Level item. Final products usually have several, what they call Indentured Levels. The more complex, can be hundreds if not thousands of levels deep, such as a commercial passenger jet or they can be only two or three levels deep, as in a simple door latch assembly. This is the foundation for all of your BOM's. You need to ensure that bill of material accuracy is at a level of 99% or better. You can do this through an audit process between Engineering and the Shop Floor personnel. We have used a form called the Engineering Change Request form (ECR). Basically anyone can fill out this form to document what they believe to be a bill of material issue and engineering is responsible for responding with a resolution within a given time frame. The document should be logged into Engineering and given a control number. Tracking login dates and any aging of this log should be reported on periodically.

Below is an example of a simplified product structure for building a computer. The computer would be considered the zero level or the finished good item that you would sell to the customer. The next row below it containing the Keyboard which would be indentured level one and the next row containing the body would be indentured level two and so on and so forth. In this case you have eight indentured levels below the zero level or finished item. The more complex the product, it usually follows, the more indentured levels you would have.

Figure 1.2 Bill Of Material BOM

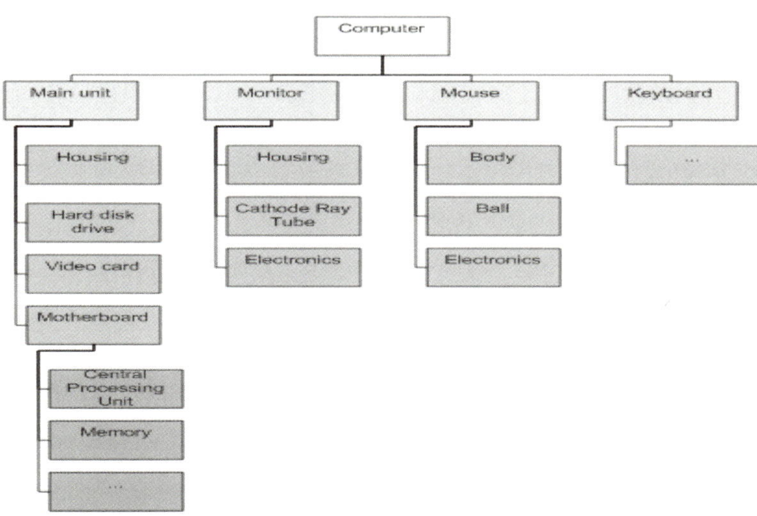

https://www.google.com/search?q=product+structure+diagram&espv=2&tbm=isch&tbo=u&source=univ&sa=X&ei=xyDPU5T_F9OkyATfw4KgBA&ved=0CBwQsAQ&biw=1920&bih=955

Back Flushing: Back flushing is a method of maintaining your inventory balances based on your reported build quantities and is often used in a LEAN environment. This is a bit more advanced management method in that it requires a very high level of bill of material accuracy of 99 percent or better. It is built on the premise that if you have reported a final assembly as complete you must have used all items required, within its bill of material (Product Structure), to do so. At this point, all items within that bill of material are decremented from inventory by part number, based on quantity per times the quantity reported complete, or back flushed.

CONCLUSION

To summarize this chapter we need to touch on some key points. The often overlooked system variables or parameters that provide the foundation for how MRP looks at your data will determine how you would like to see the data presented. This is important in getting data from MRP in a format that is acceptable by the end user of the data, assuming of course, all the other inputs are accurate. You need to understand routings and BOM's along with their respective lead time and their respective functionality in your system. Many of the modules we did an overview on in this chapter should have their own set of KPI's created for them as they are key inputs to the ERP system and accurate output is dependent on the accuracy of the data being used. The Sales forecast accuracy, MPS, shop floor performance, work order management, on time completion, accurate lead times and Bills of Material just to name a few. You need to decide what is best for your business and how best to track the level of performance on all your key inputs to your ERP system.

BRAINSTORMING FOR YOUR SUCCESS!

Use the following diagrams to brainstorm ways to improve your process and performance.

5 WHY'S

1.______________________________

2.______________________________

3.______________________________

4.______________________________

5.______________________________

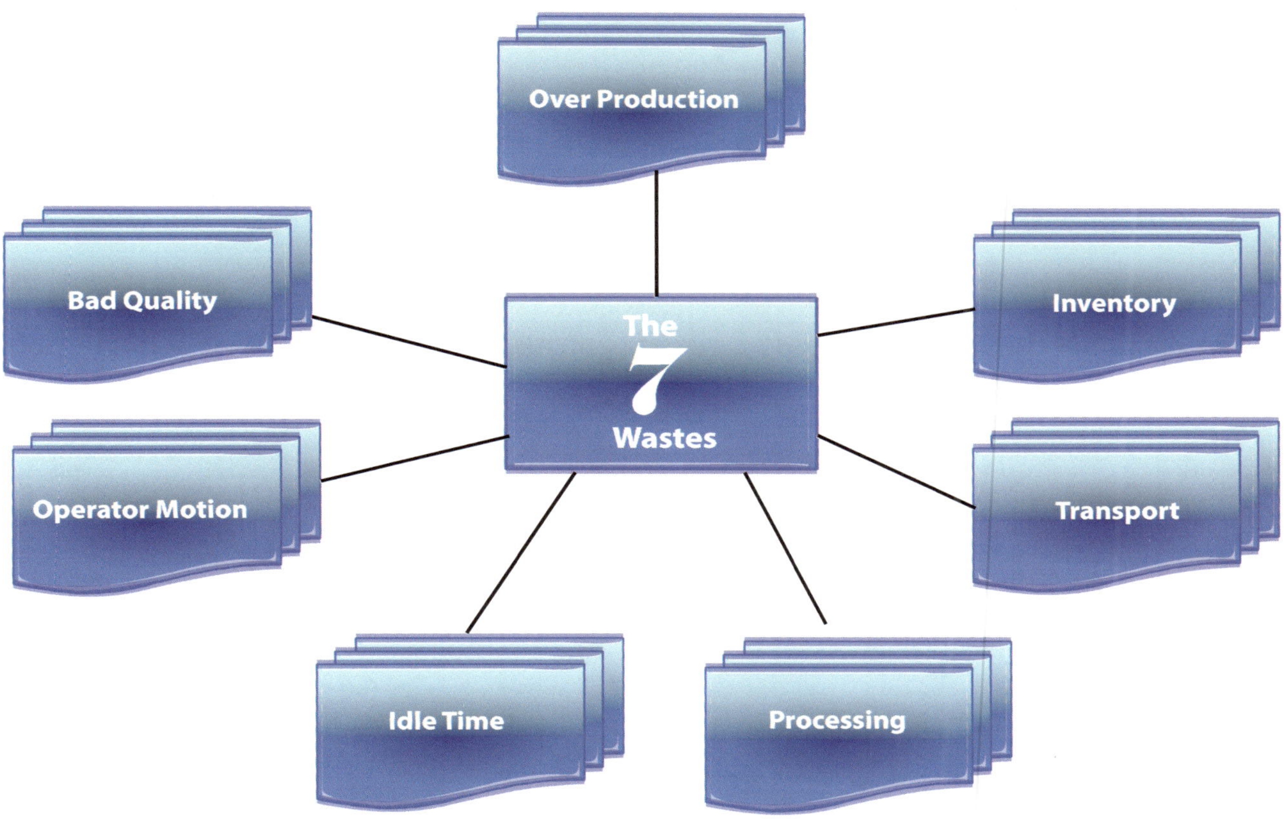

ISHAKAWA

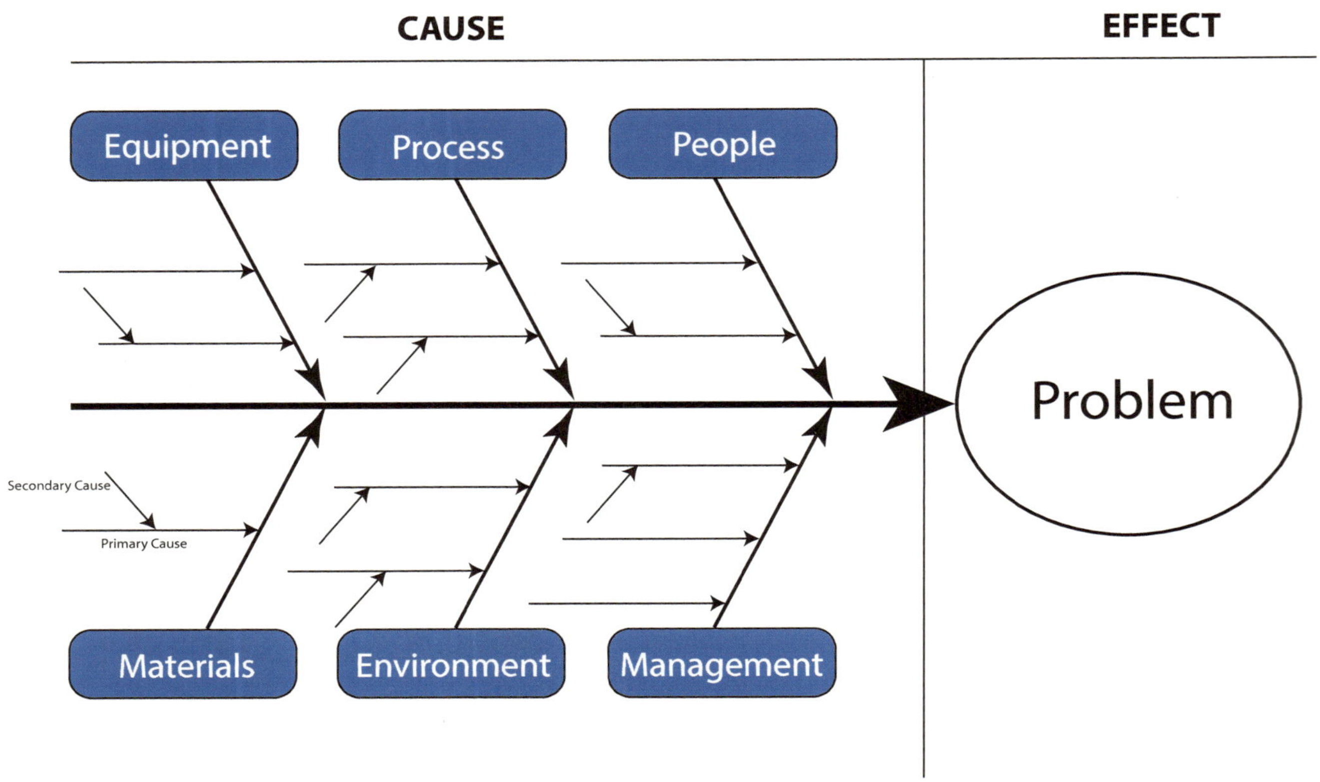
CAUSE
EFFECT
Equipment
Process
People
Problem
Secondary Cause
Primary Cause
Materials
Environment
Management

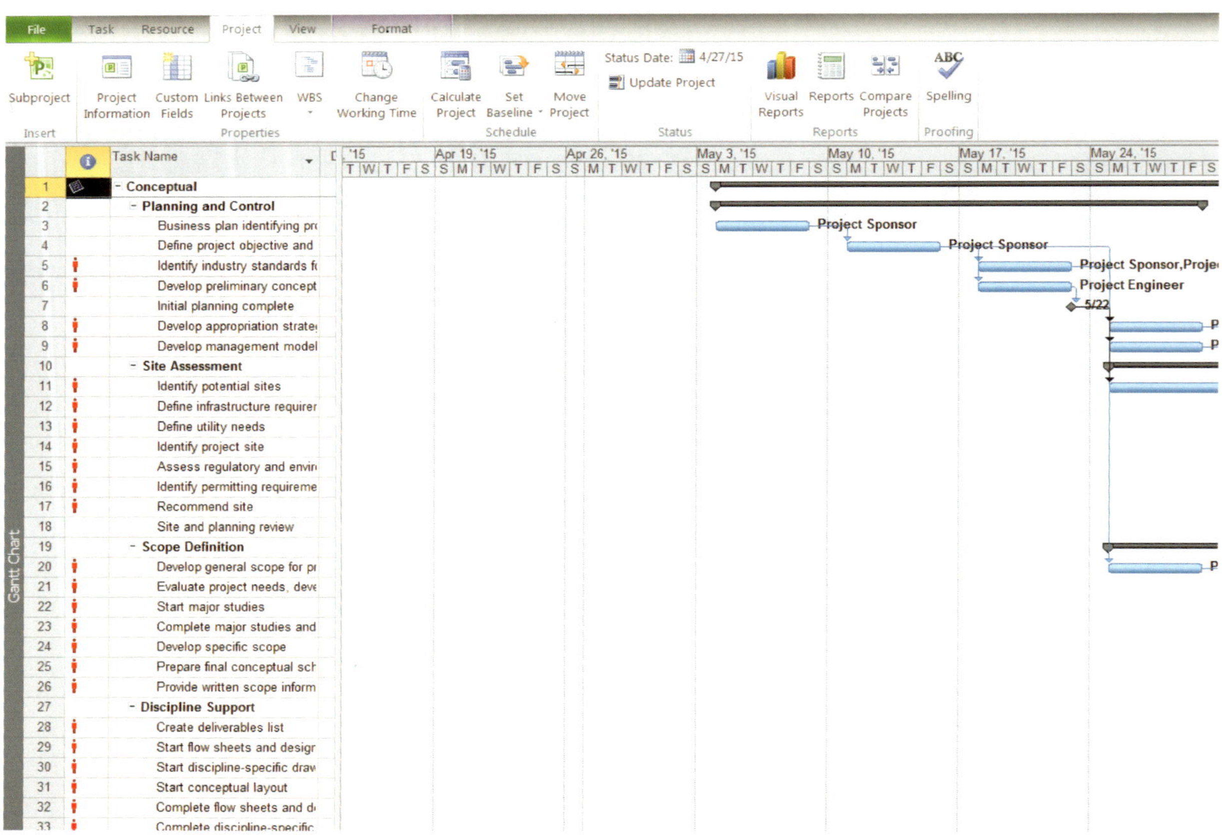
File
Task
Resource
Project
View
Format
Subproject
Project Information
Custom Fields
Links Between Projects
WBS
Change Working Time
Calculate Project
Set Baseline
Move Project
Status Date: 4/27/15
Update Project
Visual Reports
Reports
Compare Projects
Spelling
Insert
Properties
Schedule
Status
Reports
Proofing
Task Name
Conceptual
Planning and Control
Business plan identifying pro
Define project objective and
Identify industry standards fo
Develop preliminary concept
Initial planning complete
Develop appropriation strate
Develop management model
Site Assessment
Identify potential sites
Define infrastructure requirer
Define utility needs
Identify project site
Assess regulatory and enviro
Identify permitting requireme
Recommend site
Site and planning review
Scope Definition
Develop general scope for pr
Evaluate project needs, deve
Start major studies
Complete major studies and
Develop specific scope
Prepare final conceptual sch
Provide written scope inform
Discipline Support
Create deliverables list
Start flow sheets and desigr
Start discipline-specific draw
Start conceptual layout
Complete flow sheets and d
Gantt Chart
Apr 19, '15
Apr 26, '15
May 3, '15
May 10, '15
May 17, '15
May 24, '15
Project Sponsor
Project Sponsor
Project Sponsor,Proje
Project Engineer
5/22

CHAPTER 2
SETTING UP A STEERING COMMITTEE

Implementing an Enterprise Resource Planning –ERP system is not a software installation. Many think that users take the software out of the box, installs it, and magically the "Silver Bullet" is there to resolve all of the organizations issues. This is far from the truth.

The implementation of an ERP system is an evolution, from where your company is currently to where you want to be based on the project's mission statement. The mission statement is necessary to have prior to the implementation process.

An ERP project requires that there be a high level of commitment, buy-in, and involvement by upper management. There needs to be several levels of responsibility by upper management in the ERP project. These are the sponsor, the project champion, and the project manager. Included in this process are key management personnel from all the tactical groups within the organization. We must realize that the purpose of the ERP system is to close up all of the loop holes that exist and to bring the entire organization under one system and have processes, procedures, and work instructions as "Standard Work". So how does this all begin?

! The first step is to convene a steering committee that is inclusive of managers of the major tactical groups within the company. The CEO will be the sponsor since he or she will be tasking these managers to proceed with the project. At this point a Strength, Weakness, Opportunity, and Threat analysis-(SWOT) should be done. By performing this all important review the steering committee will better understand the issues. They all will be on the same page, and in time they will all come to a high level of agreement of what the Current State is, what the Ideal State should be, and how to proceed on a Future State map to get to the ideal state. Although we may never get to utopia, the purpose here is to utilize a "Kaizen-Continuous improvement process", once the new ERP system has been selected, implemented and maintained. When the SWOT analysis is performed and all of the issues are identified it is now time to solicit and select the best possible ERP solution. For now we will assume that an ERP solution has been selected. We will later in the ERP guide discuss the selection process.

! The next step is to select a project champion. This champion will be responsible for taking the Mission Statement and seeing that the final outcomes expected of the ERP implementation process are achieved. The champion will be the individual who will be the bridge between the sponsor-upper management team, and the implementation team. The champion will be the one who will take issues that the project team manager cannot resolve and get a definitive answer-solution so that the project stays within the timeline and budget. The champion will attend meetings at particular milestones in the project timeline and should report back to the upper management steering team.

! Next is where the work is performed. Not only will there be a team of employees from each tactical group within the organization but a team of subject matter experts, (SME's), from the ERP software solution provider or the ERP reseller. A project team leader from both the customer and the ERP solution provider will be working together to ensure that the mission statement and anticipated results are achieved within the agreed to budget and timeline. The ERP solution provider's team will have developed an obtainable timeline, know who will be responsible for each task required to complete the implementation and go live function, and ensure that there are measurements-metrics to know "How am I doing?". At this level in the ERP Project there needs to be a clear, correct, and concise understanding of what is going to be expected and this must be seamless as well. Once the internal team members, SMEs, and the ERP solution providers are known, the team can begin the Future State Map journey towards a successful implementation.

This is a cycle that includes a commitment throughout the organization from the top down resulting in reporting from the bottom up to management. This is a closed loop approach that is essential to ensure that all of the issues and requirements have been addressed where the project can proceed with a high level of probability of being successful. This provides a means by which to catch issues that might surface that have not been anticipated yet are addressed and resolved.

We now have a structure that will support the entire ERP implementation project but there is one issue that we have not addressed. That is "Soft Skills" and "Change Agents" required in selecting and managing the human capital on the teams. When everything is said and done, the most difficult part of the project is not the hard skills, infrastructure, software, or computer equipment. It is the human side of the project.

Whether you are buying an ERP system to replace an in-house legacy system or replacing an aging ERP system, knowing who the players are and how to manage them is crucial to a successful implementation process. This process is truly an "Evolutionary" process. Thus there will be players on your team that will be afraid of "Change" and if they resist the change they will make the implementation process more difficult than it should be. The other issue along these lines is that the SMEs that are selected for any of these teams could be resistant to change as well.

Why you may ask? SMEs have been doing what they have been doing for so long that they fail to see beyond what they know and do not have an open mind to change. A new system will take what you are currently doing and may eliminate many of the tasks, procedures, paperwork, and reporting. Employee roles and job titles will be questioned and changed to meet the new requirements of the system. When selecting team members for the implementation team, where the work is being performed, make sure you do not use just their expertise as the final benchmark for selecting your team's members. Look for open mindedness, willingness to work on teams, and ability to change one's direction at a moment's notice.

! Finally, leadership should consist of mentors that know how to deal with people. Their responsibility will be to maintain a high level of team effectiveness while mitigating issues that may arise; clearly, concisely, and correctly. The most variable resource involved in an ERP implementation project is the human factor. Good people skills and mentoring skills will be an asset providing a desirable way to develop a healthy team environment.

Many organizations develop a Mission Statement and you may see them when you walk through the front door visiting them. The same is true when a project is decided upon. Nevertheless, a mission or a mission statement by itself needs other components to make it whole and obtainable. So, how do we make this process more robust? The answer is to develop a "Charter" that supports the mission and or mission statement. What are the components of the charter?

They are as follows:

1. Mission Statement
2. Objectives
3. Scope
4. Deliverables
5. Benefits
6. Metrics

MISSION STATEMENT

The mission statement or the mission should not be a long and laborious discussion. It should be direct and define the project clearly, concisely, and correctly. The following is an example of a mission statement:

Mission: Select an ERP software package and to successfully implement that ERP solution at the ABC Aerospace Corporation. The goal is to achieve a leaner organization while improving all enterprise processes with the ability to meet future marketplace challenges. The ultimate goal is to achieve world class leadership as it relates to the global marketplace.

OBJECTIVES

With the successful implementation of an ERP solution the ABC Aerospace Corporation will be able to:

1. Reduce total cycle time and lead times
1. Improve data and data accuracy
2. Become more agile
3. Eliminate waste and become a leaner organization
4. Streamline processes and use one database and reporting system
5. Create Standard Work for all processes, policies, and work instructions
6. Meet the Return on Investment, (ROI) expectations that were stated when the project was launched
7. Better forecast future customer demands and the deliveries of these demands
8. Improve Profit Margin
9. Improve responsiveness, service and delivery

DELIVERABLES

The following can be expected deliverables by implementing an ERP solution at the ABC Aerospace Corporation:

1. Modernize the enterprise system
2. Improve collaboration with all of the tactical groups within the organization
3. Reduce and or eliminate Non Value - Added processes and reduce the time needed to do administrative functions
4. Eliminate paperwork and use workflow processes.
5. Utilize our customer portal sites so that we can meet changes in the demands requested by the customer.
6. Improve bills of materials, inventory accuracy, routing files, vendor master list, material master list, and Sales and Operation Planning
7. Better use of Human Resources

SCOPE

The implementation and use of an ERP solution will involve:

1. Finance: Receivables and Accounts Payable, Cost Accounting
2. Materials: Purchasing, Planning, Inventory Control, MRPII, Master scheduling,
3. Production Control
4. Manufacturing: Manufacturing Engineering Operations, Maintenance
5. Human Resources: Job assessments, Job Descriptions, Gap Assessments, Training Needs
6. Sales and Marketing: Forecasting, New Business-Old Business, New products and services
7. Sales and Operation Planning
8. Quality Assurance and Quality Control

BENEFITS:

The benefits that will be realized include our ability to compete in the worldwide marketplace that is becoming more competitive each year. With the ability to streamline our internal as well as our external processes we will be able to meet these challenges and grow the business while improving our "Profit Margins".

CONTROL AND MEASUREMENTS-METRICS:

This will be the responsibility of the project Steering Committee. There will be a Project Time Line that will be determined by the selected ERP Solution Providers and agreed to by the ABC Aerospace Corporation.

As it can be seen, we now have a comprehensive "Charter" that is supported by the mission, objectives, deliverables, scope, benefits, and measurements. The more clearly the project is defined the better understood the goals will be and what it will take to obtain them.

CONCLUSION

An ERP Implementation is not an installation but a project that will affect everyone within the organization and those who are external resources. Having a Sponsor and a Champion is essential and they will be responsible for driving the mission to lower levels of the organization. A well-developed implementation team will be required to ensure that the system is implemented correctly; standard operating and work instructions are developed and written. Testing and retesting of these standard operating procedures will be needed to ensure that the entire process works seamlessly and flawlessly.

Organizational skills must be considered when choosing the team players. These leaders need to practice "Change Management" to ensure that the organization not only changes its processes to the new ERP system but also ensures that those who are on the team see the need for change and benefits.

Last, a Project Charter should be drafted to clearly and concisely define the Mission, Objectives, Deliverables, Scope, Benefits and Metrics.

NOTES:

CHAPTER 3
TRAINING AND EDUCATION

Evaluating the effectiveness of employee training in any organization is crucial to the success of the Company. In evaluating training methods, much thought is given to planning what is expected from the training based on the company's needs. If the training is going to be effective the needs of the participants, their management and the company must be met. Companies can perform interviews and provide a simple survey to capture much needed information in an efficient manner. Capturing this information will help determine the skills, knowledge, behavior and attitudes that employees possess and those that should be developed. There is a great need to determine what types of skills are required for all employees that will be using an ERP system in their daily functions. [1]

Usual industry training and classroom instruction are as follows:

1) Industry training, instruction, or educational courses are provided through a variety of media:
   - Personal instruction
   - Classroom instruction; interaction with a teacher over a prolonged period of time allows for information to be absorbed and analyzed.
   - Seminars; interaction with an instructor for a short intense period of time allows for efficient learning.
   - On-the-job (OJT) training; interaction with managers intensely (mentoring) on the job allows for accurate quick assessment of learning and comprehension.
   - Written instruction
   - Text books; utilizing a written resource for the formal study of a subject for decisive interpretation.
   - Manuals; a descriptive resources used for manual hands on instruction specifically geared towards explaining how to install or operate a product.
   - Guides: a precise resource that assists in the understanding towards a directed goal.
2) Computer-Based Training (CBT); self-paced information learning using a computer for a specific skill acquisition.
3) Web-Based Training (WBT); instruction delivered over the internet anywhere, anytime with anyone for any purpose.
4) Teaching via audio tapes or video tapes
5) Distance learning; delivery course material in a virtual class where learners and teachers are separated by time and distance but collaborate for learning objectives.
6) Or in using a combination of all the above. [1]

There are many reasons for evaluating training programs and most include: showing how to improve future programs, determining whether a program should be continued or dropped, and justifying the existence of the training department and or training materials.

The commercial success of ERP systems is well established, yet many companies do not have a proper training system in place that suits their needs. One reason behind this is that most companies have pressing demands that prohibit effective training execution.

Training for ERP includes planning and implementation, change management, ERP systems administration, network resource planning and an ERP programming language.

The idea is to capture pertinent functional areas critical to ERP literacy and learning for an enterprise view, including a comprehensive understanding of business processes. Then drill down into the employees' basic performance requirements such as purchasing, production control, quality, etc.

The training activities progressively flow through business functional areas in the form of dedicated modules from specific areas such as HR, sales, purchasing, manufacturing, engineering, scheduling inventory, accounting, administration and the executive area. Exposing all employees to a clear understanding of the entire enterprise in this training will benefit the company. [2]

The opportunity for employees to access each module and drill down into particular functions through training exercises on an ERP tool allows them to better understand Cross Functionality and business process more than any other training program available. Training should include some cross-functional learning.

ERP has become a necessity in today's workplace because it helps companies integrate all their real time data and processes into a unified system that helps meet demand for products on time, at the right price and cost effectively. Employees who know how to use this ERP software will have the competitive advantage and will better assist companies to meet their strategic goals, objectives and profits. [2]

There is a demand currently for employees with this ERP skillset and experience. Understanding the flow of functions in a company, the training program provides employees with effective comprehension. Industries like the aerospace, electronics, medical, government, military, and many others in service and production areas are in need of qualified candidates with these ERP skills. There is a dire call for experienced ERP professionals and incoming candidates. Companies need to be lean and effective in all their decision-making to remain competitive.

CRITICAL SUCCESS FACTORS IN ERP TRAINING AND EDUCATION

According to Carol Ptak, in her book, "ERP Tool, Techniques, and Applications for integrating the Supply Chain," in addition to the education required, training is needed to operate the software. This training is typically included directly or indirectly in the purchase price of the software. Training answers the questions of "how." Software training will teach the people using the system the keystrokes and transactions required to run a company, thereby enabling individuals with the kind of training needed to be successful. Ptak further cautions against putting an untrained individual on a live ERP system because of the high probability that this system will crash if this course of action is taken. Without the understanding of the tool's integrated approach, the implementation will fail. [3]

Many ERP implementation experts, like Ptak, stress the importance of proper education and training for both employees and managers. Most people will naturally resist changing the way they do their jobs. Many analysts have noted that active top management support is crucial for successful acceptance and implementation of such companywide changes. [2] Monk and Wagner go on to say that some companies willingly part with funds for software and hardware, but don't properly budget for employee training. ERP software can be complex and intimidating; this fact supports the case for adequate training. Gartner research recommends allocating 17% of the projects budget for training. Those companies spending less than 13% on training are three times more likely to have problems. [3]

Some of the critical factors in ERP training and education involve:

- staff that do not know the system yet can be trained
- management that participates in training and encourages all employees
- following through with ERP training
- an ERP database designed for training
- money for the training
- precise functional instruction
- cross-functional collaboration for the ERP training.

These factors are important to success, as ERP allows for easier global integration in a time when doing business globally is increasing significant. The system removes barriers to currency exchange rates and other pertinent international issues.

Accordingly, ERP systems allow managers to manage operations not just monitor them. An ERP system can dramatically reduce costs and improve operational efficiencies.

As stated by Carol Ptak in her book; "ERP Tools, Techniques, and Applications for Integrating the Supply Chain", most ERP implementations fail because the education component is neglected. [3]

The other critical issue in ERP training is data record accuracy, and employees not knowing how to properly input data. If employees do not trust the information in the system, they will seek out other ways to do their jobs letting this very expensive software sit underutilized. (Excel Spreadsheets).

To have effective ERP implementation, accurate data is required for inventory records, bills of material, (BOMs), routings, sales orders, work orders, purchase orders and all execution transactions. The quality of the information is directly dependent on the accuracy of the data entered. Training of employees in the use of ERP is a critical element to a company's profitability, growth and success.

SELECTING THE RIGHT NEW EMPLOYEES

Companies using ERP must carefully consider choosing the right internal and or external people for ERP functional positions; those who have the right skills, the potential for comprehending and the agility to think outside their current activities. Internal employees who learn the system to teach others should be SME's in their company process, procedures, products, and service. This repository of experts will help sustain training. By having this group of experts, most issues will be quickly addressed and resolved.

Lack of understanding of what is required in the new positions responsibility will inhibit the potential for bringing in the right candidate. Criteria for the required position should therefore be established and those new candidates should be able to exhibit ability to comply or perform activities illustrating their ERP knowledge and execution.

ERP TRAINING AND EDUCATING NEW HIRES

A company must be able to have the resources, tools and knowledgeable employees in place to quickly and accurately assess their business needs and customer considerations. The need for controlling costs while successfully satisfying customer's needs is essential. The influx of changes in technology now allow for a broader scope of opportunities in international telecommunications and internet activities. Companies are in business 24/7. Learning how to manage this information and manipulate this software is an important priority for most businesses today. Their future results depend on them putting all the necessary processes in place including training. New employee orientation should include ERP training as it pertains to that new employee's functional responsibilities for which they were hired.

Many companies strongly value employees ERP skill sets, attributing their successful operations to their knowledgeable ERP employees. The argument is significant in expressing that; organizations that use ERP, without a doubt need employees that are ERP savvy to be a successful entity. There have been many cases of companies that have failed due to lack of training of their employees.

ERP TRAINING AND EDUCATION OF EXISTING EMPLOYEES

Training and updating employees on ERP is a major challenge. People are one of the hidden costs of ERP implementation. Without proper training, about 30 percent to 40 percent of front-line workers will not be able to handle the demands of the new system. The people using the keyboard are now making important decisions about buying and selling -- important commitments of the company. They need to understand how their data affects the rest of company. Some of the decisions front-line people make with an ERP system were the responsibility of a manager earlier. It is important for managers to understand this change in their job and encourage the front-line people to be able to make those decisions themselves. They own that and they are empowered to make those decisions.

Training employees on ERP is not as simple as Excel training. ERP systems are extremely complex and demand rigorous training. It is difficult for trainers or consultants to pass on the knowledge to the employees in a short period of time. This "knowledge transfer" gets hard if the employees lack computer literacy or have computer phobias. [3]

In addition to being taught ERP technology, the employees now have to be taught new responsibilities. With ERP systems, one is continuously being trained. Companies should provide opportunities to enhance the skills of the employees by providing training opportunities on a continuous basis to meet the changing needs of the business and employees.

There appear to be critical issues in the ERP training in industry due to time, money, resources, individual managers not being qualified or knowledgeable enough to teach, mentor or engage in training because of the serious nature of their other functional responsibilities.

MEASURE OF ERP TRAINING AND EDUCATION SUCCESS

Research shows that all users of the ERP system should be properly trained in using the system to get the maximum benefit from it. Human Resources should perform a Gap Analysis or skill set matrix and then decide on what Interventions will close the gaps. [3] Employees in different functional areas will have different training needs from one another. Management must decipher information for decision-making while staff will need to know how to use the system to complete their jobs. All employees should be trained and educated in the ERP basics, the overview of the functional areas, the cross-functional relationships, and how actions in one area affects, other areas. [3] Impact of ERP Training and Education in Workforce.

The human element is considered one of the most valuable assets of a corporation. Individuals must know what is required of them and how to perform those tasks. It is wise to mention that any ERP training and education program, strategy, or initiative, must have management approval and have top down direction. Without management attention this critical tool will be used as a word processing device instead of the integral resource of which it is capable. This is why identifying patterns and standards in industry relative to documentation, procedures or policies, will prove to help sustain the ERP objectives.

Business professionals have gained significant advantage from ERP training and education by becoming more useful on the job due to the exposure of the material. Companies need to prepare and train employees to be as useful as possible, especially those without prior training. Research shows the benefits achieved through training and education as part of change management activities. ERP training can result in certain benefits for the company, as the employee will be suited better for his/her job.

Most organizations have main functional areas of operation: Marketing and Sales (M/S) Supply Chain Management (SCM), Operations, Accounting and Finance (A/F) and Human Resources (HR). Each area is comprised of a variety of narrower business functions, which are activities specific to that functional area of operation.

! The primary focus of ERP is to improve and increase the flow of accurate information within an organization, in order to make better business decisions. ERP software improves cooperation, collaboration and interaction between all business tactical groups.

CONCLUSION

It is the intent of chapter to convey and expose information in ERP training and education. The ERP training and preparation for employees that are new to ERP, or have experience using a different ERP system comes down to creating effective training practices resulting in skilled knowledgeable workers. The impact on the workforce of a skilled worker is crucial for success, growth and improvements. An ERP knowledgeable worker adds immensely to the productivity of the corporation.

NOTES:

NOTES:

CHAPTER 4

KPI'S—
CURRENT STATE OF THE COMPANY ASSESSMENT

Key Performance Indicators (KPI) measure the business health in the enterprise and ensures that all individuals at all levels are going in the same direction towards the same goals and strategies. [5] Any Dashboards created uses the chosen KPI's to reflect a visual representation of the specific KPI's to understand the Current State. KPI provides a vehicle for standardization, collaboration, cooperation and coordination across the enterprise. [5].

KPI's help to get insight in your business performance -- "What gets measured, gets managed". KPI's are also known as performance metrics, business indicators, and performance ratios. Performance measures that are established for a process, and/or steps, of that process.

Metrics is a mathematical function, a verifiable measure, that associates a real nonnegative number, either quantitative or qualitative terms defined with respect to a reference point. Companies need metrics to function efficiently and effectively daily, weekly, monthly and yearly. [2]

DEFINE CRITERIA TO DO ASSESSMENTS

Managing operations of a company requires that executives, managers, and employees need to plan for controlling resources used in the process of manufacturing a product, servicing a part or providing a service. Resources required to do this could include; labor, materials or capital. The flow of material up and down the supply chain involves many variables. The control of this process and the need to identify those elements that are essential to capture in data gathering. Decisions need to be made constantly and having up to date accurate information about these key indicators will allow the company to be successful. Selection of key performance indicators can vary by a company's current state need. They are not the same for every company but generally there are some that are constant to the profitability of any company. These include financial, external and internal.

Management must assess performance for potential improvement if necessary. How else would they determine the effectiveness of the process? Companies will consider that they absolutely need to measure and then evaluate their company's current and past performance. They absolutely need to identify their future goals per their corporate strategies

There are a number of questions relative to KPI issues, they are:

WHAT SHOULD BE MEASURED?

- How many metrics should you have?
- How often should you measure?
- Who is accountable for the metric?
- What should you use as a benchmark?
- How do you know the metric reflects strategic drivers, mission, goals and vision of the company?

KEY PERFORMANCE INDICATORS (KPI)

KPIs, as identified by local manufacturers, and as found in literature, include some KPI as follows:

- OEE (Overall Equipment Efficiency)
- Manufacturing Schedule Compliance / Control Compliance
- Improved Customer Service Performance and Response Time
- Employee Performance
- Lean Initiatives
    - Reduction of Waste
    - Reduction in Operating Costs
    - Production Planning Improvements
    - On time delivery (Complete Shipments) Order Fulfillment Rate
    - Reduction in Inventory levels (Inventory Accuracy)
    - Utilization of Resources
    - Increased Profit/Revenue
    - Reduced Time to Decision
- Inventory Carrying Costs
- Inventory Turns RMA cycle time

EVALUATE WHAT YOU DO WELL, NOT SO WELL OR NOT AT ALL

Measuring or monitoring supply chain performance reveals the gap between planning and execution and helps companies to identify potential problems and areas for improvement. However, it is recognized that developing key performance indicators (KPIs), or metrics, is very challenging and a set of practical guidelines is not readily available for companies and supply chain management (SCM) practitioners. [3]

The experience from, and the review of, industry standards and best practices in supply chain performance measurement suggest that "less is better" as to developing performance metrics. Companies should focus on only a small list of KPIs which are critical for their operations management, customer service, and financial viability. Potential KPIs should be developed for each of the supply chain operations-reference (SCOR) model's four meta- processes (plan, source, make, and delivery) and need to be hierarchically grouped such as primary and secondary metrics. [3]

Despite the importance of performance measurement for successful SCM, many companies wonder how to put performance metrics in place and make them work in practice.

DETERMINE YOUR BUSINESS DRIVERS

BUSINESS DRIVERS

- Reduce Costs
- Improve Customer Experience
- Manage Growth Expectations
- Improve Customer Response Time
- Interoperability Issues Across Multiple Operating Locations
- Identifying personnel responsible
- Streamline and accelerate process to improve efficiency and productivity

OPTIMIZE THE USE OF CURRENT CAPACITY

Provide visibility to business processes across functions and departments and modernize technology infrastructure and applications.

WHAT DATA TO COLLECT AND FREQUENCY OF REPORTING

The data that will be needed to be collected will be determined by what drives the company and what their key performance indicators are. Each companies supply chain must adopt a more balanced perspective in its performance measurement and management. Companies need to recognize the importance of the drivers of strategic future performance. Managing a given supply chain's overall performance necessitates the coordination of measures across the different entities on the supply chain, often requiring all entities to adopt a common balanced perspective in their performance management to facilitate the overall performance and competitiveness of the entire supply chain.

CONCLUSION

Performance measures and metrics are essential for effectively managing operations, particularly in a competitive global economy. The global economy is featured with global operations, outsourcing and supply chain and e- commerce. The real challenge for managers of this new enterprise environment is to develop suitable performance measures and metrics to make right decisions that would contribute to an improved organizational competitiveness. Now the question is whether traditional performance measures can be used and out of them which ones should be given priority for measuring the performance in a new enterprise environment. Some of the traditional measures and metrics may not be suitable for the new environment wherein many activities are not easily identifiable. Measuring intangibles and nonfinancial performance measures pose the greater challenge in the so-called knowledge economy. Nevertheless, measuring them is so critical for the successful operations of companies in this environment. [5]

NOTES:

CHAPTER 5
WRITING PROCEDURES

You are now the owner of an ERP solution and the teams have been created. The ERP reseller has developed an implementation plan with your internal steering committee. And the ERP implementation team manager is ready to proceed with the implementation plan-process. As the teams start to learn how to navigate the ERP system, they will also be required to draft their Standard Operating Procedures and there will be Work Instructions that will become "Standard Work" for anyone who uses the ERP system.

Earlier it was stated that this is an evolution and it will transform your company on how it operates and how work is done throughout the entire organization. In addition, it will also change the way you collaborate with outside sources, the customer, and others that are part of your supply chain. Thus it is essential to have accurate documentation that states how the work gets done. Meaning, no matter who is doing a process in any of the tactical groups, there is a standard way on how the work is performed.

Why is this important? Each tactical group, division, and department within the organization has specific duties to perform each and every day. When work is performed it needs "Input" and "Outputs". Information is received, reviewed, and then next course of action is taken to complete the next step of the process. If there are any "Gaps" then the process cannot be brought to closure and then the inputs cannot be translated into outputs for the next process to begin. To ensure that this does not occur, these gaps must be closed and the way to do this is with a well-defined Standard Operating Procedure-SOP and Work Instructions.

Let's look at a typical manufacturing company. Where does it all begin? What is the workflow? How do we check to see that the work performed is yielding the anticipated outcomes?

Marketing and Sales first and foremost responsibility is to bring in new sales and to see what the customer needs as far as new products. A quote master would be generated. If this is successful a new customer order is booked. Then there is a need to meet a schedule it will then be determined how much will be required throughout the life of the contract. Here the input is defined by the contract details added to the backlog. The outputs are those that are going downstream to the other tactical groups within the internal and external supply chain.

At this point, a Sales and Operation Plan needs to be developed to see if there are any issues where the Operations side of the business cannot meet the demand of the Sales side of the business. Once this has been satisfied the orders are released to Engineering, Operations, Materials, Quality Assurance and other tactical groups. They then will start their processes and once all of these groups have developed the way these products will be manufactured, purchased, inspected, assembled and tested the work orders and purchase orders are then released.

As we can see, there are a lot of processes that need to be performed in order to get a product to the customer. This cannot be done in a haphazard fashion and must be structured. When the orders from contracts are released to be reviewed by a Sales and Operations review, how is that done? When there is agreement on the "What and How" how does this get communicated to engineering, operations, quality assurance, materials, and other tactical groups within the organization? This is done through documentation, and procedures, that define the process and the work instructions. This is Standard Work and must work with upstream and downstream processes. In other words, this must be a closed loop system! If this is not a closed loop system, the hand-offs will not be translatable for the next user-customer and the process will come to a stop. The purpose of an Enterprise Resource Planning System is to have all of the inputs and outputs work ensuring that required tasks-work is completed accurately and on time. More importantly, we wear the customer's hat when we receive information and then wear a supplier's hat when we send information to our next customer whether they are internal or external in nature.

Remember this, "Bad processes beat up good people"! If the process and work instructions that you are using doesn't satisfy your customer, then there needs to be a review of these processes and work instructions. This is to ensure that they work properly and that they meet your customers' needs and requirements. Again these are both internal as well as external customers and suppliers.

What is a Procedure?

A fixed, step-by-step sequence of activities, or course of action (with definite start and end points) must be followed in the same order to correctly perform a task. Repetitive procedures are called routines.

To take this one step further if you are an ISO certified company these procedures and work instructions must be in your ISO manuals. The acronym ISO stands for International Organization for Standardization. This organization has been in force for the past twenty plus years. In order to do business with companies around the world you must be ISO certified. ISO certification is broken down into many ISO standards. Some of these are ISO 9000-Quality Management, ISO 9001 is the same as ISO 9000 but is for companies who design products. ISO 14000 is utilized to meet environmental standards, and ISO 22000 is for Safety Management Systems. All of these have one thing in common, "What you say is what you do". The AS9000 standard is utilized for the Aerospace industry and was developed by the major airframe manufacturers and what is true for the ISO standards mentioned here is the same for the AS standard.

STANDARD OPERATING PROCEDURE TEMPLATE

The following items are part of a procedure:

1. **Who performs what action(s)?**
2. **What sequence they perform the steps in the task.**
3. **The criteria (standard) they must meet.**

Figure 5.1 Standard Operating Procedure Template

ABC Company Inc	xxxx Department	SOP #
	xxxx Divison / Function	Revision #
		Implementation Date
Page #	• 1 of	Last Reviewed/ Update Date
SOP Owner		Approval-May requrie more than one signature depending the affected departments

1. Purpose

- Describe the process for <Name of SOP>.
- Describe all relevant objectives for this procedure.

2. Scope

- Identify the intended department/s and work activities where the SOP may be relevant.

3. Prerequisites

- Outline information required before proceeding with the listed procedure;
- For example, worksheets, documents, work instruction overlapping procedures, etc.

4. Responsibilities

- Identify the personnel by tittle and department that will have a primary role in the SOP and describe how their responsibilities relate to this SOP.
- Include contact information where applicable.

5. Procedure

- Provide the steps required to perform this procedure (who, what, when, where, why, how).
- Add a visual where applicable like a process flowchart.

6. References

- List all applicable resources that may be needed when performing the procedure;
- For example, Admin policies, Municipal Code, government standards, work instructions and other SOPs.

7. Definitions

- Provide a definition for procedure or process specific terms or acronyms.
- Provide other sources as needed to help better understand this SOP.

Figure 5.1 is an example of an ISO acceptable format for a procedure.

This is why procedures and work instructions are so important

Your procedures (along with your ISO 9001 quality manual and required forms) make up your Quality Management System (QMS). Your procedures will describe how you operate and control your business and meet the ISO 9001 requirements. Thus when you change your "Procedures" your ISO manuals will also reflect these changes. When you are audited by your ISO Registrar they will be looking at these procedures-work instructions and then go out to where the work is performed and ask to see if you are in fact following them (Objective Evidence). If you are and the procedures are relevant and up to date then your ERP system should be running smoothly which will reap the payback that you anticipated in the justification for purchasing it. Now let us turn our attention to "Work Instructions". They go hand in hand with Procedures and ultimately with the company's ISO manuals. So what is a work instruction?

Work Instruction: A work instruction describes how to perform a task, which is a stand-alone document of the procedure such as "Completing a PO" or "Ordering supplies".

You may need more detail than that described in the procedures. Many businesses include work instructions to aid in training, to reduce mistakes, a point of reference for jobs, etc. In an ERP system there will be many work instructions such as:

1. How to split a work order
2. How to enter an order into the backlog
3. How to generate a requisition and purchase order
4. How to create a Bill of Materials
5. How to develop and Parts Master List or a Vendor Master List
6. How to record Labor and the Quantity of parts completed by an employee
7. How to requisition inventory- parts and sub-assemblies to Kits for assembly and testing

! All of these Procedures and Work Instructions must be proven out during the implementation process of the ERP system. They must be TESTED OVER and OVER again to ensure that there is continuity between the different stakeholders within the ERP system and that there are no GAPS within the system. Once each procedure and work instruction is created and proven out, the next step is to "Hand-Off" your outputs to the next customer downstream. If they do not have any issues with your outputs then their inputs will be accurate and then the process starts all over again for their customer! In the end, if everything works seamlessly and accurately we will have created a "Closed Loop System"! This is the ultimate goal!

Once these Procedures and Work Instructions are proven out, they should be written on your company's stationery (letterhead) and there should be a cover sheet where those who must approve these documents can sign and the date of the approval will in the applicable space provided. The signatures here are for all of those departments who are affected by these documents. This sign-off sheet will be the first page of the Procedure and the Work Instruction. Sometimes Standard Operating Procedures-SOPs are used in lieu of Procedures.

BE AWARE Nevertheless, they must be tested and approved prior to it becoming "Standard Work".

Last, your ISO manuals will then have all of these documents; Procedures and Work Instructions. The registrar will then audit your company and if you are following them then you will maintain your ISO certification. But remember, whenever you modify-amend your procedures and work instructions you MUST update your ISO manuals to reflect the latest ones being utilized by your company. In this case "Ignorance is not a good defense"! So maintain all of them and perform "Internal Audits" on a cyclical basis!

CONCLUSION

An ERP implementation is an evolutionary process. An ERP system requires that all of the end users develop, write, and utilize Procedures and Work Instructions. These documents are the centerpiece of an ISO Quality Manual and "What you say is what you do"! Procedures must work when your outputs are handed off to your customers downstream. If they work then the customer's inputs and outputs to the next customer will work as well. Make sure that your ISO manuals reflect the latest revisions of your Procedures and Work Instructions. When you are audited by your assigned ISO registrar he or she will be asking you to show them objective evidence that you are using them and that you can demonstrate they are being used at the time of the audit. You should perform internal audits on a cyclical basis and make sure that any revision and or amendments to your Procedures and Work Instructions are updated in your ISO Quality Manuals.

NOTES:

CHAPTER 6
ASSESSING DATA INTEGRITY

EXISTING SYSTEM REVIEW

Assuming you have some sort of system in place now, MRP/ERP, a manual work order system or some sort of platform you are currently running your business on, you will need to do Data Mapping. Data Mapping is the process of matching data sets in your current software to the equivalent data sets in the new software.

This process will involve the entire Steering committee. The purpose is to define what data fields in your current system are equivalent to what data fields are in the new ERP software. Nomenclature as it pertains to field headings differ from product to product within the wide range of ERP software currently out in the marketplace. Some Field Names may even be "Country of Origin" specific. However, although data sets may have different names the data is the same. For example, the data migration may be from your current ABC field in your accounting module to the XYZ field in your new software's accounting module. If it isn't obvious already it is vitally important that this be done correctly and be agreed upon by all those that may be a user of this data either currently or in the new software environment.

Now, to be clear, data mapping is an arduous task and is often facilitated by an Information Technology person and the head of Accounting. However, all disciplines must be represented. The IT profession will actually perform the mapping and the migration of the data once origins and destination fields are defined. This is usually done through a scripting process best explained by an IT professional, not us. The data should always be first migrated into a test database that we spoke of in an earlier chapter. The data should be migrated and tested via whatever reporting you will require insuring the mapping process worked effectively, accurately and that 100% of the data migrated.

NEW ERP SOFTWARE ASSESSMENT AND SELECTION

As you may have already guessed, there is a myriad of software out there and wading through all of the pros and cons of each package is a daunting task at best. Before you even begin the review and selection process you must first decide what the critical must haves for your company are. At the highest levels what is your core product/s? Are you a Manufacturing Company, Pharmaceutical Company or a Service Company?

Once you decide, you can do much of your research on-line. Try to narrow the field down to about a half dozen software packages that claim expertise in the areas you have decided that need to be your focus for your company. You can contact each in turn and request a demo disc if they have this available or many have online website that can be used for testing and training.

In addition, it is critical to get a list of references from each. This is probably the single most important input you can get. With this list of references you can select the references that most closely compares with your current business. Hopefully, you will end up with at least three that are a close comparison to your business model.

This activity should be assigned, as a task, to a group formed from the Steering Committee. Naturally all findings need to be documented for each software package investigated, because this will need to be presented to the Steering Committee at a later date when final selection is necessary. Contact all the references you received from the software company. Although, you will only get references from them that they know will be positive, you can still learn quite a bit from this exercise.

Discuss the strengths and weaknesses found by the user of the software you are considering. Often it will be helpful to create a matrix of key requirements with them delineated across the top and the software companies listed down the left hand side.

Table 6.1 ERP Software Comparison

ERP SOFTWARE CANDIDATES	COMPANY				
	A	B	C	D	E
Attributes					
Accounting	X	X	X	X	X
Mobile Apps		X	X		X
Serves Small Businesses	X	X	X	X	X
Serves Medium to Large Businesses		X	X	X	
Reporting and Analytics	X	X	X	X	
24/7 Customer support	X		X	X	X
Onsite support	X		X	X	
Financial Management	X	X	X	X	
Add modules as needed for custom solutions	X	X	X	X	X
Cloud capable	X	X	X	X	X
Cost	**$250,000**	**$245,000**	**$275,000**	**$265,000**	**$195,000**

The above table is an example list of elements to consider when choosing an ERP system. It can be used to review each candidate, to compare and contrast the attributes of each to best suit your needs. This matrix, along with other key elements under consideration, like support, costs, upgrades and modular expandability are all key elements needed to make an intelligent software selection. Make sure you enlist the help of all departments in this phase and ensure you get a list of attributes from each department represented on the Steering Committee and those departments that may not have a representative on the committee. Request a list of "'Must Haves" and a list of "Like to Haves" from each department head.

Garbage in Garbage out (GIGO)

! As simple a concept as this may be it is always astounding to us how often higher levels in the organization just don't get it. The software is just a tool, it will not fix those things that you are doing wrong or poorly. This bears repeating: The software is just a tool, it will not fix those things that you are doing wrong or poorly. It is critical, as we outlined in a previous chapter, that your processes, procedures and work instructions support the need for accurate and timely data. If you have not done this, there is no point in mapping or migrating bad data to the new software. As a matter of fact, we would definitely go as far as saying that the cleanup required in chapter four could eliminate the need for a software upgrade even temporarily, if you were thinking that new software was going to solve your inaccurate data output problem to begin with. It will not!

Establishing percent levels of accuracy for those critical inputs to ERP

We will attempt here to delineate a list of KPI's that you need to have in place and their relative level of accuracy. Keep in mind this list is by no means exhaustive and you will need to determine their fit and other KPI's that may be unique to your business model.

1. **Inventory Accuracy:** This refers to your active parts and should reflect an accuracy level of 98% or better, through the proper use of cycle counting and A, B and C coding.

2. **Bill of Material Accuracy:** This needs to be maintained at a 99% level of accuracy or better. Keep in mind this is the driving force for material purchases and down-line work order creation. This includes operation sequence routing, standard set up and run times along with labor standards. This will be used for shop loading and labor management along with identifying productivity and efficiency on machines or labor intensive work centers. This is an effective audit process that must be put in place and routinely monitored and reported on.

3. **Productivity:** Smply stated is a formula: Efficiency X Utilization = Productivity. The goal here is always 100%. However, you must determine what level of productivity is acceptable for your situation. You should initially target no less than 80%.

4. **Work Order Material Variances:** Should be no greater than 5%. You should track this on two fronts; first material usage variance and second material dollar variance. This will help in keeping your usage rates accurate and your cost of goods accurate.

5. **Work Order Labor Variances:** This will help you maintain accurate labor requirements, this in turn along with accurate Bills of Material will make projections against Rough Cut Capacity Planning an accurate tool. Labor variances of greater than 5% need to be investigated and resolved.

6. **Shop Floor Execution:** The on time completion of work orders will support lead time for on time delivery to the customer. This is of course fairly reliant on the on time release of the work orders, the BOM and the routing methods accuracy. A simple tracking methodology is planned completion date to actual completion date. Initially you should set a goal of 85%.

7. **Buying and Planning Activities:** If at all possible and it usually is, you need to introduce the Buyer/Planner concept to your organization. We have implemented this concept three times in our experience/career. The Buyer/Planner is a much more effective asset to the company. Knowledge of both disciplines creates the understanding that allows this position to be more cost effective and timely in their ability to

execute. You can track this by tracking several attributes, the on time released of work orders, which is planned vs. actual and the conversion and aging of planned orders being turned into actual work order or purchase orders. You can set an aging period of say seven days from today. You should be set up to track this by Buyer/Planner and report on a daily basis.

8. Forecast Accuracy: We know, forecasts by definition are inherently wrong. While that may be true, the more astute among us, and now this includes you, want to know how wrong they are and why. This should be tracked on several fronts, one by the salesperson, two by product line and three by region. You'll be surprised how the accuracy begins to increase when you track, post and report on this essential KPI's. In addition, knowing the level of the inaccuracy, by salesperson, by product and by region is an effective planning tool in and of itself. When you begin to define trends, as you gather history, these trends can be built into your Rough Cut Capacity Planning (RCCP) methodology. The basic tracking methodology is generally looking at Forecast vs. Bookings. Begin to track this and with the input of those in the trenches begin to assign accuracy goals, but don't allow them to be anything less than aggressive in nature.

9. Supplier Performance: Lead Times play a critical role in this performance measurement, as does Quality. These lead times should be maintained by the Buyer or Buyer/Planner (depending on your organizational structure) each time an order is placed for a part. The system stated lead time needs to be validated. The supplier should not take a hit on their performance if we place an order in violation of their stated lead time for the part. This supplier performance measurement is a result of tracking a supplier's on time delivery and their delivered product quality. This is often a weighted average of 60/40 in favor of quality. This supplier report card should be done monthly and the results reported to your supplier. The goal is 95% or better. This will be tiered results with your preferred suppliers at the 95% or better level and then you set up lower level percent thresholds for a good supplier, fair supplier or a poor supplier. All poor and fair suppliers need to be worked with closely to improve their performance or you must define an exit strategy. Within the ERP system, there is a Master Vendor Approved List that must be updated to address these issues.

10. Customer on Time Delivery: Well, this is the ultimate goal. You want to have 100% on time delivery to your customer. Heck, your customer wants you to deliver 100% on time! This is the goal that kind of tells you how you are doing with all your other performance measurements. The fact of the matter is, as you demonstrate improvement in all the other KPI's we have mentioned here, this KPI will inevitably improve. Once again, you as a company need to decide that whatever the demands from the customer are it is your responsibility to meet those demands. That means even if they place an order in violation of your lead time you do not have the luxury of not taking an on time delivery hit.

CONCLUSION

IDEAS You must always strive to be the "Best in Class" for your product offering. This means your stated lead times, although they currently are accurate, and they should be, must be challenged and you need to find ways to reduce lead time to the customer. This is the path of continuous improvement. The never ending need to be more efficient, effective and have reduced lead times to your customer. Wherever that goal leads you is where the next challenge lies in getting a little bit better tomorrow.

NOTES:

CHAPTER 7
OK, WE DID OUR ASSESSMENT, NOW WHAT?

DEFINE CRITICAL ELEMENTS OF DATA TO SUPPORT INTEGRITY AND OUTPUT

As it was discussed in the prior chapter, accuracy of the data that is going to be utilized in the new ERP system is critical! Why you may ask? Because it is required data that is utilized throughout the organization. If modules in an ERP system share the database and there are inaccuracies within the database, the error is then multiplied many times rendering the system to be useless. The expectation of process improvements and enhancing the bottom line will be compromised.

Another issue that needs to be considered is the strength and knowledge of your internal information technology-IT group. They are the ones who are going to be involved with other subject matter experts-SME's to help in the migration of data from your legacy system to the new ERP system. In chapter 6 it was stressed that if you input garbage in you have garbage coming out (GIGO) of your newly implemented very expensive ERP system.

The next element here is how much assistance you are going to depend on from the software reseller and or the ERP solution provider. In Chapter 6, Table 6.1, we show a comparison of choices of features of attributes that can be considered in a planned budget. There should be a rough cut budget initially based on the features and support that you want including labor. Meaning, dedicated time spent by your employees. You need to identify internet and intranet accessibility and requirements needed. If there is a weak IT group involved internally and there are no services being provide by either the ERP reseller and or the ERP solution providers, the task ahead will be very difficult and very expensive while ensuring that the project timeline will be longer than you would anticipate. So the TEAM needs to be in place to make sure the data can be brought over and that it is accurate!

Now let's discuss the legacy data. Most legacy systems have the essential building blocks to run an enterprise but they lack a highly functional relational database. If the company is going to be dependent on utilizing the new ERP system, then the database has to be relational. Meaning, all of the data is related in some way and can be accessed across the enterprise. Bills of Materials, Master Material Lists, Master Vendor Lists, Routing Files, Inventory Files, Labor, Attendance systems, Contract, Contract Bookings, Work Orders, Purchase Orders, and MRP are some of the building blocks of a well-defined relational database! If the labor, work order, routing files, run and setup times are not correct, then this string will be broken and the data shared will be of no value. If this happens, you will then hear the end user state; why did we make this transition when the legacy system was better? In reality, it wasn't better, you just found ways around the data to get it right. In our discussion, about Lean Manufacturing, we discuss several rules.

They are all essentially used in the ERP Implementation and data migration processes:

1. Structure every activity
2. Connect Customers and Suppliers
3. Specify and simplify the flow path
4. Experiment at the lowest level towards the ideal state. (1)

Then REFLECT on the outcomes!

Why are these lean pointers so important? An ERP system is your global support system of all the company's data, both inputs and outputs. This includes processes that are created to enter the data into the database, receiving data as a customer and sending data as a supplier. If the ERP system is not valid then we can say there is going to be a lot of confusion and distrust of the database and of its outputs to the end-users.

There are the *Seven Wastes of Lean*: **TIMWOOD**

- **T**ransportation
- **I**nventory
- **M**otion
- **W**aiting
- **O**verproduction
- **O**ver processing
- **D**efects

DEFECTS

The ones that apply here are: Over Processing, Defects, Inventory of Bad Data, and Waiting for the data to be corrected. This must not happen and the database created must be correct and tested over and over again to ensure that we do not see these wastes as a result of the database's inaccuracies! To do this, a team must be involved from the outside sources (reseller or ERP solution provider) as well as the internal IT group-department. In addition, there will be SMEs from the tactical groups within the organization that will be asked if the data looks correct and when errors are found, how these errors can be corrected. Remember, the SMEs and their associates are going to be dependent on the newly created database. So it is in their best interest to get it right. Ultimately, a "Total Team" effort needs to be established to see this through to a successful end.

Write procedures to support and ensure data accuracy

As it was discussed in chapter 5, procedures and their accuracy is of the utmost importance. Without the use of "Standard Work" as is used in policies and procedures, the entire process is like a "Rope of Sand". A rope of sand is very hard to pull on or push. In an implementation process of a new ERP system, it is important that the implementation process be as precise, accurate, and that there is an open line of communications to resolve issues, policies, and procedures quickly. They then need to be retested to see if they mitigated the issues encountered. This is true when developing the database and all tactical procedures and work instructions!

Who should be involved in developing procedures to ensure the data is accurate? This is a hand-shake between the external team members as well as the internal team members. The external team members will need to receive data from your legacy system so that it can be migrated to YOUR new ERP solution. In turn, your internal team members will have to know what the legacy database contains and how they will setup the files in a utility program to get the data from your legacy system to YOUR new ERP system. This should all have been defined when these two groups initially met and understood the importance of the: who, what, where, when, and how the processes will be utilized for the data migration.

Data fields are probably the most important. Let's look at the levels in the Bill of Material-BOM; if it is Make or Buy how many per in each level, material used; specifications ASTM, Mil, J Specification, etc. If the fields coming from your legacy system are migrated to the new ERP system and they do not line up then the database is not usable. You should start out with a small sampling of data to see how this will all work out. It is much easier to use a small database and then make the necessary changes required. If your legacy system's database is consistent, then the issues found when corrected should allow the entire database for the BOM to be migrated with little or no issues as a result.

If your legacy database is in the form of Excel spreadsheets then the process will be more difficult and thus this would be as if you are starting out to create a new database to be used across the entire enterprise (In our experience we used Access). For now, we will assume that the data is coming from a structured database within your legacy system.

If we are looking at marketing and sales, accounting, engineering, manufacturing engineering, planning-MRP, procurement, time and attendance, maintenance, and other tactical groups all of these groups have a database that is related to their function yet ALSO share databases from the other tactical groups. This is truly a "Relational Database"! Once the procedures and work instructions are generated and tested, they will be completely dependent on the overall "Relational Database". If the data is wrong - it is wrong across the entire enterprise! So the development and proving out of the database is a priority and a crucial step before one can continue with the implementation of the ERP solution!

! To do the aforementioned, procedures need to be written so that the proper steps can be carried out to ensure when the data is migrated and tested that it is valid. This is a TEAM effort by both the internal and external team members of the data migration process. Once the data migration has commenced, there may be a need to modify these procedures to accommodate changes in the process. Meetings between all of those who are involved should take place on a cyclical basis. The objectives of these meetings are to analyze the current state, ensure the outcomes are what is expected, and identify what needs to be done, is being done, to resolve outstanding issues. Teamwork and collaboration is of the utmost importance and by all means TEST, TEST and RETEST until it is correct and has met the implementations KPI's and timeline requirements!

TEST ALL PROCEDURES

This is quite self-explanatory. The fourth rule of Lean as mentioned before states "Experiment at the lowest possible level" then reflect on the outcomes. In other words, this is "Continuous Improvement" at its very best! The reflection here is to understand what the outcomes are and then if it is determined that they are not acceptable, decide what needs to be done to improve the outcomes of the procedures. This will ultimately determine the outputs of the system.

The database team, inclusive of SMEs from the tactical groups within the enterprise who are responsible for the database accuracy must challenge the database over and over again. There must be a high level of agreement that the database is accurate and it will be able to be utilized by each tactical group within the enterprise to develop there processes.

Metrics need to be established, (KPI's as mentioned in chapter 4), to see where you are currently, (to create a baseline), and then review how you are going to improve the performance and accuracy of the database's inputs as well as the outputs. Please note one size does not fit all. There may be a need to have some custom "Scripts" written in order to migrate your legacy's database to the ERP solution's database. This should not be the RULE but the EXCEPTION. Writing custom scripts are costly and they may not work when you upgrade the ERP solution to the next revision level!

Testing is the only way to ensure that you have satisfied all of the requirements for accurate data. This will insure that the downstream processes will have fewer issues to deal with later on in the ERP implementation process! Remember, "Measure twice, and cut once"! (2) The course of action to take is to test, test, and test. This is much cheaper to do it this way then to send inaccurate data downstream to your customers both internal and external.

CONCLUSION

The database of an ERP solution is the heart of the system. You will need to have a team of qualified experts, SME's assigned to this project. The transfer of your legacy system's database will need to be reviewed before it is migrated to the new ERP solution. This must be able to be defined so that it will be transferable via a utility program (Access) from the legacy system to the ERP solution's database. Within this activity, continuous improvement is a must!

Process procedures will need to be written and tested to ensure that they work correctly allowing the process to continue with little or no issues. This is a "TEAM" effort and everyone will need to be held accountable for their tasks that need to be carried out. Meetings should be setup on a cyclical basis so that there is an open line of communications to discuss, review, and resolve issues as they relate to database accuracy. SMEs are a critical factor since they are going to be using the relational database that comes out of this process. Their inputs are critical.

Again test, test, and test again until you get it right. The only way to meet your expectations, to keep costs down within budget, to meet the project timeline is to ensure the accuracy of the database. This is in your best interest to do so.

NOTES:

CHAPTER 8
CREATING A PROJECT PLAN

Creating a project plan is essential for keeping the project on track and within budget. The keeper of the project plan should be whomever is chairing the steering committee and she or he should be responsible for keeping it updated. The entire Steering Committee will have input to the plan and the current status of the individual projects. The project plan is and should be a living document. By this, we mean it should be updated at each meeting. It can be added to as necessary when new issues make themselves known.

NOTE: Remember to add notes to each task identified in the project plan. This is where all the detail will be kept concerning each project. This is where you break-down the subsets of each project, if necessary, and provide some guidelines for a successful task completion.

Timelines can be adjusted and discussed when it becomes apparent we may not hit the original target date. When a task is slipping against its projected scheduled completion date it will become apparent at the weekly Steering committee meetings. The project plan will make this divergence from the plan known to the committee sooner rather than later and adjustments can be made to get back on schedule.

Many companies will have existing project management software with which they can begin to build the project. Assuming you have someone who is well versed in its use, you can start to build a task list. If you don't have someone who is adept at its use, get someone trained. If you don't have a project management software tool you can go online and search for "Free Project Management Software."

Early on you will need to brainstorm with the committee members about the list of tasks associated with the implementation. Do not worry about getting the list 100% complete, if you try to you will never get started. You should be able to identify about 80% of the total project in this first pass. As you identify tasks, you must also identify the task team leader and the members needed as part of the task team. Get buy-in from each task team for the estimated duration of the task and any special requirements, whether it be some external support, training or input from the chosen software support team. It is important to note that the software support team is an excellent external resource that you should be able to use to off-load some of the implementation workload, depending on the software support package you purchased. Do not hesitate to use them as a resource and to avoid re-inventing the wheel. They may have access to certain data that will save you time and money. Always keep them well informed of your project status and look for their input. You may be pleasantly surprised at what they can bring to the table for you in terms of implementation experience and general software knowledge.

Download a ***Gantt Chart :*** *www.ganttproject.biz/download.*

The best way to approach a list is chronologically. Think about your process end to end and start at the beginning. The beginning for your company may be sales and the end may be after sales service. Keep in mind that some projects can be done in parallel while others will be sequential. In other words, a task may require a previous task be completed before it can begin. It is important that these tasks be identified as such and set up in the project plan in a logical format. Basically, there are certain tasks that will require another task as a prerequisite. It is much like a college course prerequisite, like taking Accounting 101 prior to Accounting 201. These prerequisites are critical to the success of all subsequent tasks that are built on it. If you miss the scheduled completion of a prerequisite task you will jeopardize all down-line task completions and this may have a snowball effect and delay the entire project. Should this occur, you will need to try and make up the lost time by accelerating subsequent tasks. You may be able to accomplish this by adding additional horsepower, more people, to subsequent tasks or invoking the help of the software support team.

Once you have a list defined, with the associated leader and task teams assigned, you will need to identify those tasks that are considered to be "Critical Path Tasks." We will call these (CPT's). What is a CPT and how do we know which tasks they are? Often, what gets defined as a CPT is dependent on your business model. What we can do is give you some guidelines to help you to define and ferret out, which tasks you should consider to be a CPT. What is a critical path task? Microsoft Office Project 2003 defines critical path tasks in the following manner: "What ultimately determines the length of my project?" The answer is the critical path, which is the series of tasks (or even a single task) that dictates the calculated finish date of the project in Project path and the resources assigned to them will help ensure that your project finishes on time". That is as good a definition as any we have seen. In simpler terms, it is a task or series of tasks that if not completed on time will delay the completion of the entire project. It is built on the precept that if you take care of the pennies, the dollars will take care of themselves.

As the committee chair if you manage the critical path items to their successful, on time, completion then you would have completed the project on time. However, it is incumbent upon you to know the overall status of all the tasks within the project and work with the team to keep them all on track. By the time you have listed all your projects with time lines and defined those that need to be sequential to others you will have defined the overall length of the project. Keeping in mind of course this is not an open ended issue and you may have time constraints that will limit the length of time you have to complete the project. Once you have defined your project in this way you need to compare this to the time allotted to the project. If by some miracle they match you're good to go. Chances are you will have less time allotted to complete the project then your project plan dictates. Remember, you still have a company to run. Depending on the gap of time that exists between the project plan and your allotted time to complete you will need to make some adjustments. Start with your "Critical Path Tasks." By throttling those, you can manage the actual completion date of the project to fit the time allotted.

How to manage a gap that you can't bridge? If you have a task that you can't bridge, meaning you can't compress your timeline to fit the allotted time, you need to get creative. Once you have exhausted the internal sources for additional horsepower and the external sources for help like the Software Implementation team then you need to break down the overall task into logical subsets. You need to identify the critical path items within the subset of tasks that are required to complete the overall task that will allow you to move forward. You need to be able to set certain tasks aside temporarily that are not crucial to the overall completion of the task and circle back to these items once completed. Keep in mind though you cannot sacrifice the integrity or the quality of that tasks completion. Remember it is better to miss a deadline and beg forgiveness then it is to have a failed or ineffective project.

Once you feel you have a good project plan in place it will need to be used as a performance measurement and as a gauge for where you are overall against the completion of the project. This is the tool you can use to keep upper management informed of the plans progress and identify any bottlenecks along the way. The project plan that is a part of this reading is an overview. The elements in this plan are all critical path tasks. These tasks will generate, in many cases, their own project plan, as a subset to the overall plan.

Figure 8.1 Gantt Chart Example

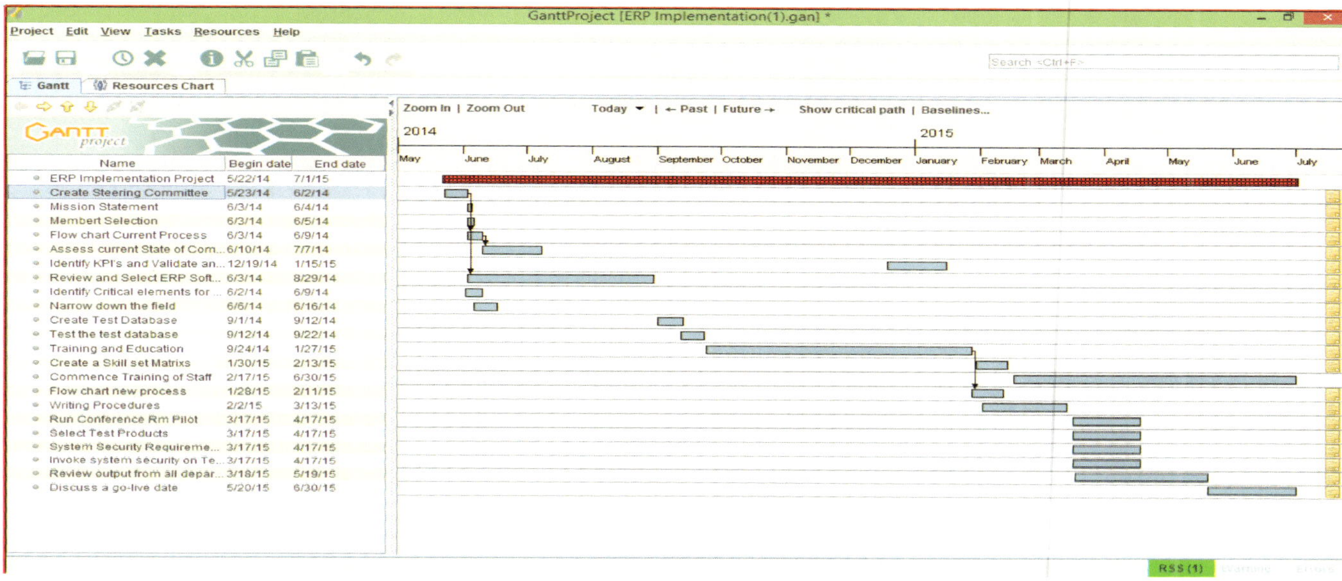

CONCLUSION

During the course of this guide we have made an effort to try and emphasize those tasks that are of particular importance to the overall success of the implementation project. You may have already noted that there are many aspects to a project of this magnitude that can adversely affect the success of the project. The project plan itself is one of those elements whose importance cannot be overstated. It is, after all, the tool you will use from the beginning of the plan to its successful completion. It is, in fact, your only concrete gauge of the plans progress and your only early warning system should things start to go off track. You must use it effectively and the project team needs to become one with the project plan. This is a tool that must be used daily by all members of the project team and all should have intimate knowledge of where there project are with the overall plan. If used effectively the Project Plan can ensure the on-time and on-budget completion of your ERP Implementation.

NOTES:

CHAPTER 9
CONFERENCE ROOM PILOT

This is perhaps one of the most important phases of the project in that it brings together:

1. The knowledge of the workflows
2. The basic functional flow
3. The players involved

Learning of the new ERP system functions to develop a unique set of processes for the company.

During this pilot each functional area representative is expected to use the new ERP system to develop new processes using the system functions. This is a very repetitive process in that some decisions are then tested and then revised as more information is developed. This pilot phase should last between 30 to 60 days and requires solid commitment from all members of the teams. If decisions regarding company policy are required then they are brought to the Steering Committee for fast action. It is critical that this phase be seen as a true catalyst for change and that persons involve stay creative and think 'outside the box'. Too often the teams revert back to their current procedures rather than explore new methods of doing things.

System Pilot Testing

After the Process Owners have worked individually and in teams to create procedures, the entire 'end-to-end' process must be tested. This is done with all Process Owners present. A well-constructed Pilot Test will have a 'test script' which is a list of all of the primary functions that the company must be able to perform to process their work. For example; this can start as early as in the quotation process and end in the transactions that are created being posted to the financial modules. This would include areas such as sales, engineering, production planning, inventory, etc. Each step in the test script is given a 'pass/fail' mark. A step that failed must be researched and corrected. The test is performed as many times as is required to have all areas pass.

During the steps, up to this point, the project has relied on the Process Owners to develop and test these new procedures. Typically, the Process Owners also speak with the Process Implementers in their areas during the pilot process to verify facts or get new ideas. Once the process Owner has completed the Pilot Testing, they are now ready to fully involve the Process Implementers. This team is comprised of the persons that will be responsible for using the new procedures every day. During this step, they are brought into a training room and the new procedures are presented to them in a 'hands-on' training environment.

BE AWARE

When a system 'Goes Live' it means that the new procedures are now in 'production'. It is critical to make sure that the new procedures will actually support the business. The purpose of this phase is to develop a test plan that will put the system through its paces and hopefully prove that the procedures will work when the system goes live.

Years ago, it was common to test the new procedures by 'going parallel'. This meant that all of the company's business activity would be done both under the old system (legacy) and then the new system – essentially running two systems at once. Today this has become increasingly difficult and is now considered near impossible to do by most mid-sized companies. To meet the requirement of testing, however, this phase develops a companywide test plan that is less than a full parallel test but still meets the requirements of proving that the system has a solid chance of being successful.

The methodology that is used is based on testing a subset of the transactions for a period of time. For example, selecting five to ten customers whose orders will be processed in parallel for a period of 30 days might be a good test plan for a company. Each test scenario will be different depending upon the size and type of company.

CONCLUSION

In summary, the importance of running a conference room pilot cannot be overstated. This is the company's opportunity to exercise all of the processes it takes to run an organization from booking an order through to after sale service. It is essential that there is complete participation by the entire cross functional team and that all inputs and outputs are validated along with ensuring seamless hand-offs from one department to the next. This is not where you want to take any short cuts because finding flaws in execution or process issues here is far more expedient and much less costly then after you go live. If you do a good and thorough job in this phase, going live should be painless and a positive experience for the entire organization.

NOTES:

CHAPTER 10
ERP MEETINGS

Meetings should be weekly in an attempt to get the status and assign action items of all activities pertaining to this project. Upper management should have already approved this, be fully supportive and have a senior representative in attendance. Having meetings will allow for a better understanding of what has been completed, what needs to be completed, by whom and within what time frame.

This allows for a retrospective discussions about how, what, why, when and any issues related to completion items as well as how they can be improved as the project continues. The meetings should consist of a representative from each functional area to reflect that functions needs within the best interests of the project. The steering committee should have a charter outlining the entire projected outcome by time and these should filter down to each team created. In the initial implementation, the Steering Committee, the Process Owners and the Process Implementers should be involved.

The Steering committee should provide the vision, expected goals, with deliverables and outcomes desired while establishing timelines, providing the funds, and making decisions required to move the project along. Should the teams get stalled, the Steering Committee will provide continued support. This will create accountability and sustain the project.

The backbone of the implementation will be the process owners, who will perform the tasks and run the team meeting with a representative from each major department (function) within the company.

The Process Implementers will complete the steps determined to define how the new system (ERP) will be used. This team is responsible for using the system and testing for the "Go Live" tests.

There should be a chair and someone taking minutes for each meeting. The team chairs will be meeting separately with the steering committee. The committee should follow rules such as Roberts Rules of Order or any previously established guidelines. The meetings should be short and concise with actions determined to be accomplished. The purpose is to discuss what has transpired and what needs to be accomplished within the next weeks time frame per the milestones of the project.

IDEAS If possible, have a stand up meeting to forego any monopoly of time in this meeting. Consider just reporting at this meeting since planning has been established. All discussion will have to be consistent with the Agenda previously provided.

Team creations and responsibilities will have to be worked out prior to regular weekly meetings in efforts to have the right people on the right teams at the right time doing the right thing during the project. All members of the team should be on time to avoid wasting time and resources. The Steering committee will determine which task teams are needed and who should be on these teams with clear guidelines and objectives to be met.

The agenda will be specific to the team's goals and mission. Each team's purpose weekly will be outlined in their specific agenda which forces objectives to be met.

Initially the goals of the implementation will need to be determined. The Project teams sets up a Project plan. The project plan will hopefully have a current operational flow of the business, the team will do a complete workflow review to understand the major information flows in the company.

At this point, it is very important to match the functions of the software selected with the workflows identified.

In the plan, the Process Owners should provide a schedule for training on the functions of the software for all affected employees. The plan should include a Pilot, time for data conversion and preproduction testing.

The meetings for this implementation and the successful use of the ERP System chosen should include measurable metrics, (KPI's). These need to be clearly defined, measured and accounted for as compared to the expectations and goals. Setting up these practical, and objective goals at the start of the implementation will help in measuring the outcomes. Use current KPI's to assist in the preliminary planning.

Project leaders need to understand that there are clearly defined goals and documentation for accomplishing this implementation. There might be a need for additional resources or even new considerations during this process. Meetings will enable visibility and communication for an effective implementation.

CONCLUSION

Weekly meetings are needed to update the steering committee on progress that is being made and there should be a representative from upper management in attendance. They can provide support for the steering committee. There should also be functional area representatives on the steering committee.

To begin, the ERP implementation, the Process Owners should be involved. The steering committee provides the vision of the implementation process. Upper management representatives will be needed to resolve impasses that may occur during the implementation process of the new ERP system.

The Process Owners are the backbone of the implementation process. They are responsible for how the implementation will proceed, how it will work, and when testing of processes and procedures will be needed.

Minutes should be taken at each meeting and Robert's Rules of Order should be followed for an orderly and productive meeting. Tasks should be assigned and all of the team members should be on time. The project team sets up the project plan and there should be a schedule for all training that is needed. Piloting of processes and procedures must be included in the project plan to ensure that they work.

Metrics need to be established at the start of the implementation and there should be specific KPIs identified. These are necessary to ensure that the project plan is being carried out and that the results meet the expectations of the ERP implementation plan. The project team leaders will clearly define the goals and documentation for accomplishing the ERP implementation.

Other resources may be required which were not identified initially when the implementation process started.

NOTES:

CHAPTER 11

FLOWCHARTING YOUR CURRENT AND FUTURE STATE

FUNDAMENTALS OF FLOWCHARTING

Here we will get into a little more detail on flowcharting. Last chapter was an overview of how to approach your first meeting and how to best prepare for it. Flowcharting your Current State will help you to take the next step and that is to identify where you are today, the fundamentals of flowcharting and what you are hoping to accomplish with this process.

There are many free software websites for flowchart software. You're welcome to review Google "Flowchart Software for Free" and see what you can come up with and go with the one you feel most comfortable with. The one we have used is Lucidchart.com and we will be using this website for the illustrations that will accompany this section of the book.

The symbols used for flowcharting are universal in meaning so that no matter what software you choose the symbols will represent the same thing to the reader.

Table 11.1 Flowchart Universal Symbols and their meaning:

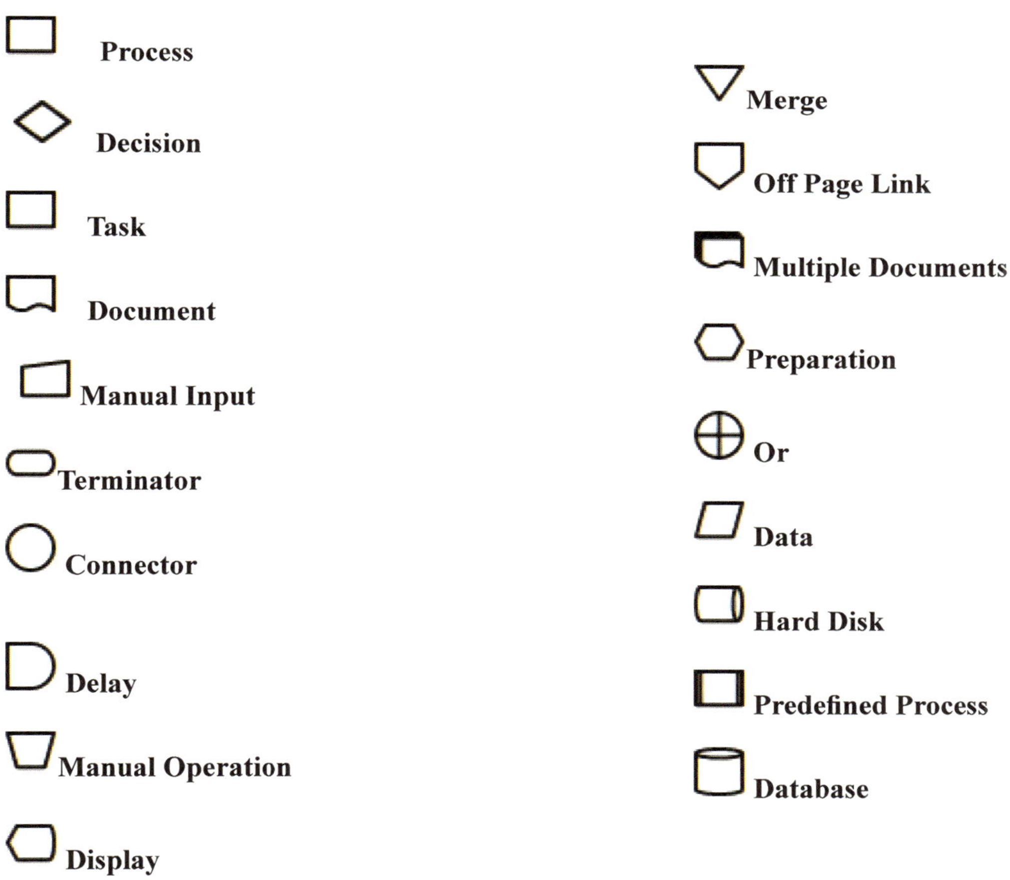

The above symbols represent the most commonly used symbols for flowcharting. Some of the names of the symbols may be obvious in their meaning while others, not so much. You will find each of the symbol names noted above in the glossary of terms at the end of this book should you need further clarification.

The best approach to flowcharting, for the purpose of ERP Implementation and Identifying where you're company currently is, is to start at the beginning of your company's processes. In many cases that is upon the receipt of a customer's order.

At the initial Steering Committee meeting, brainstorm your business process with the entire team. Be sure to write down the beginning of the process. For our purposes here, we will say it is the receipt of a customer order. Also, include any forecasting that sales does and how they arrived at their numbers. Your role as the facilitator is to keep the conversation moving forward. You need to keep saying "OK, what happens next" and wait for the feedback from the group and write down the responses. You should be doing this on a large white board or large paper and easel will also work. Have the committee's Scribe help you with this, as it is possible you may miss some input. While documenting the process flow, make sure you stop at any point where a hand-off occurs.

After the meeting you need to sit with the scribe and organize all of the input you received and using the flowcharting software put that portion of the process together. Make sure you identify the person or department that is responsible. You need to identify them, preferably by title rather than by name, because names can change and you want the eventual procedure to still be valid even though someone may leave or move to another position. The flowchart should be the visual representation of the procedure. The procedure however will have more detail in it. Once you feel you have flushed out that portion of the process, be sure to review this with the rest of the committee. You may need to tweak it a bit more as a result of this review with the committee. You are not going to do this for each step in the process. You are going to use this as a learning experience for the committee and they will each take their own portion of the process.

First, however, you must outline the entire process end to end with the committee, at a high level. You are simply going to identify key points in the process and each key point will be assigned to a committee member. The details along with the flowchart and procedure will come from them. Having gone through it the first time you should be able to assist the other team members as needed. Make sure you keep the software resellers implementation team involved. You will get much needed input from them on how these processes will work in the new software. They should be at as many of these meetings as possible and should be copied on the minutes of all the meetings. Flowcharts and procedures should also be copied to the new software implementation team. They can often suggest tweaks to a process that would help to make it more compatible to the new software.

You need to be the first one through the gate, so to speak, with anything that may be new to you and/or the committee, such as flowcharting. Learn the process well, involve the implementation team from your new software group and be sure to engage the entire Steering committee. Have yourself and the scribe take a lot of notes and document everything from the meetings and publish the minutes. Be sure to copy upper management and the new software implementation team. Constantly ask for feedback and input from all involved parties. Document all outcomes and ensure that your procedures support your flowcharts and vice versa. At the end, you will end up with a complete end to end process flowchart with the associated procedures. Have the team buy off on all the completed tasks before moving on to the next phase.

IDEAS Below is the flowchart for a generic Sales Process. You process will no doubt differ, but this will give you an idea of what you would be looking to do.

Figure 11.1 Sales Flowchart-New Business (B2B)

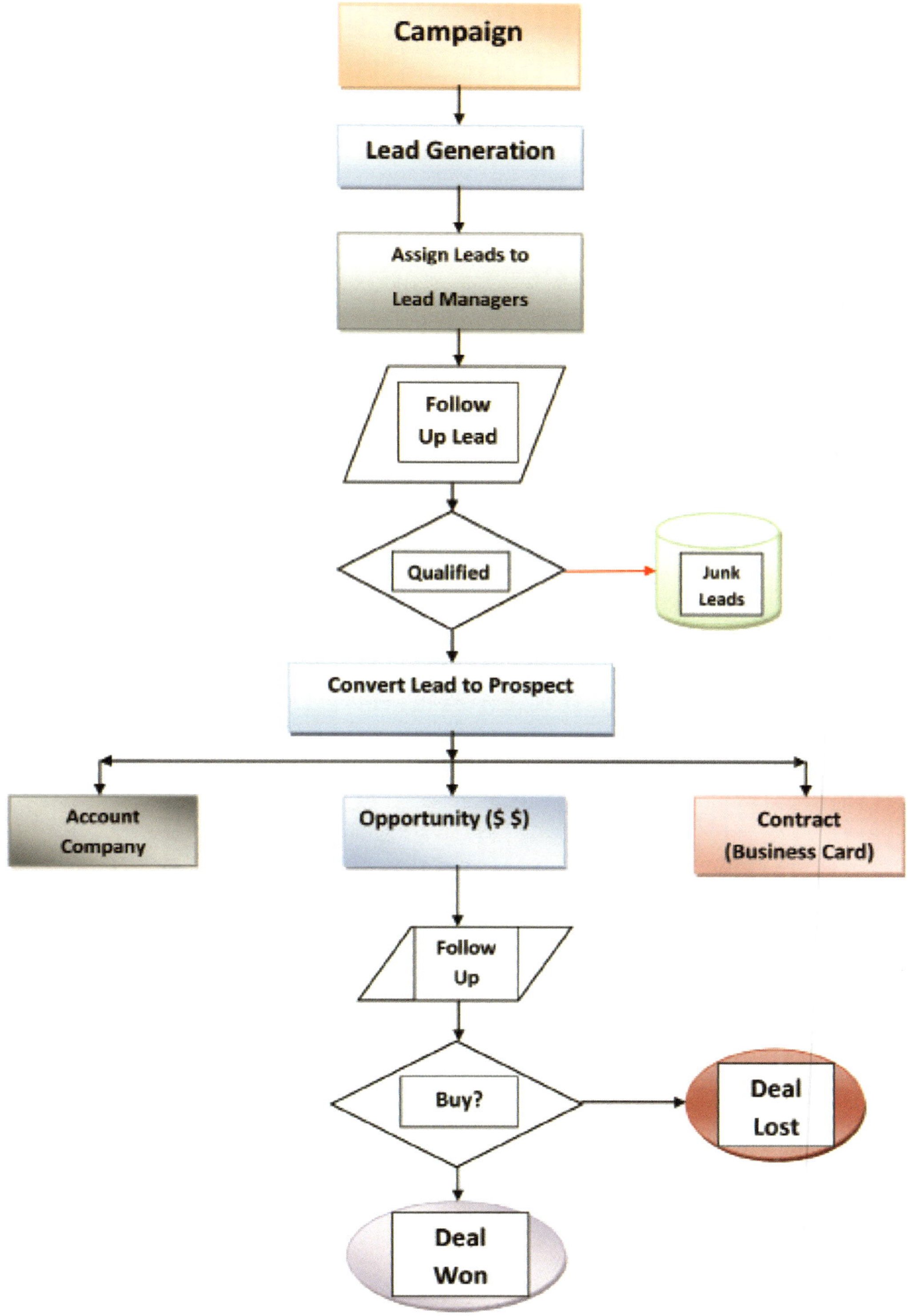

EFFECTIVE USE OF FLOWCHARTING, AND HOW TO USE THEM IN YOUR IMPLEMENTATION PROCESS

Flowcharting the Current State

So you selected the ERP system, gathered together your team players, have a charter as to what is to be expected so now let's do the implementation. Right, well not exactly! How many of those who are involved know what the Current State is of the current legacy system?

The purpose of investing in a new ERP system is to naturally replace the existing legacy system so that you can operate more efficiently and effectively. Understanding the Current State is of the utmost importance.

What does this mean? This is a technique used to review and observe the current practices and processes so that we can understand where the process works and doesn't work. It also asks if this is going to be migrated over into the implementation process and become the standard in which work will be performed going forward.

If you were to ask several people in your organization on how a requisition is initiated and what the steps are in the requisition cycle at your company, we bet you would get several versions of this process. This begs the questions; is the Process robust enough to be used going forward and does it make sense to use the processes in the ERP system being implemented?

Review

We have processes that we are currently utilizing to perform work to complete a task. We may or may not know if the Current State is effective and will yield the anticipated results we are expecting when we Go Live with the new ERP system. So how do we do this? We utilize a Current State Flow Chart Process to identify all the parts of the process and to also recognize waste in the process. Why is this important? Why would we want to duplicate what we currently have, if the process is broken, fix it! If the process has a lot of waste, eliminate it!

Next, does the ERP system have a process that is used within the scope of the ERP system? If it does then we will have to understand what the delta is between the current process and what can be expected in the new process. Change will occur and the entire process will be transformed to the new desired process. Remember, it is paramount to keep an open mind when going through an implementation process. The advantage here is to first, understand the limitations of the current legacy processes and then compare them to the similar processes in the new ERP system. There may be no need to utilize the old legacy processes and go right ahead and implement the new processes.

When you are first going through the implementation process, all of the current legacy processes should be mapped. Why you may ask? This will allow all of those in the implementation process to understand clearly what it is, how it is done, and what can be expected when the process is completed. It will also affect both upstream and downstream processes and hinder their performance if the process is not robust or has problems incorporated in them.

Figure 11.2 Example of Flowcharts

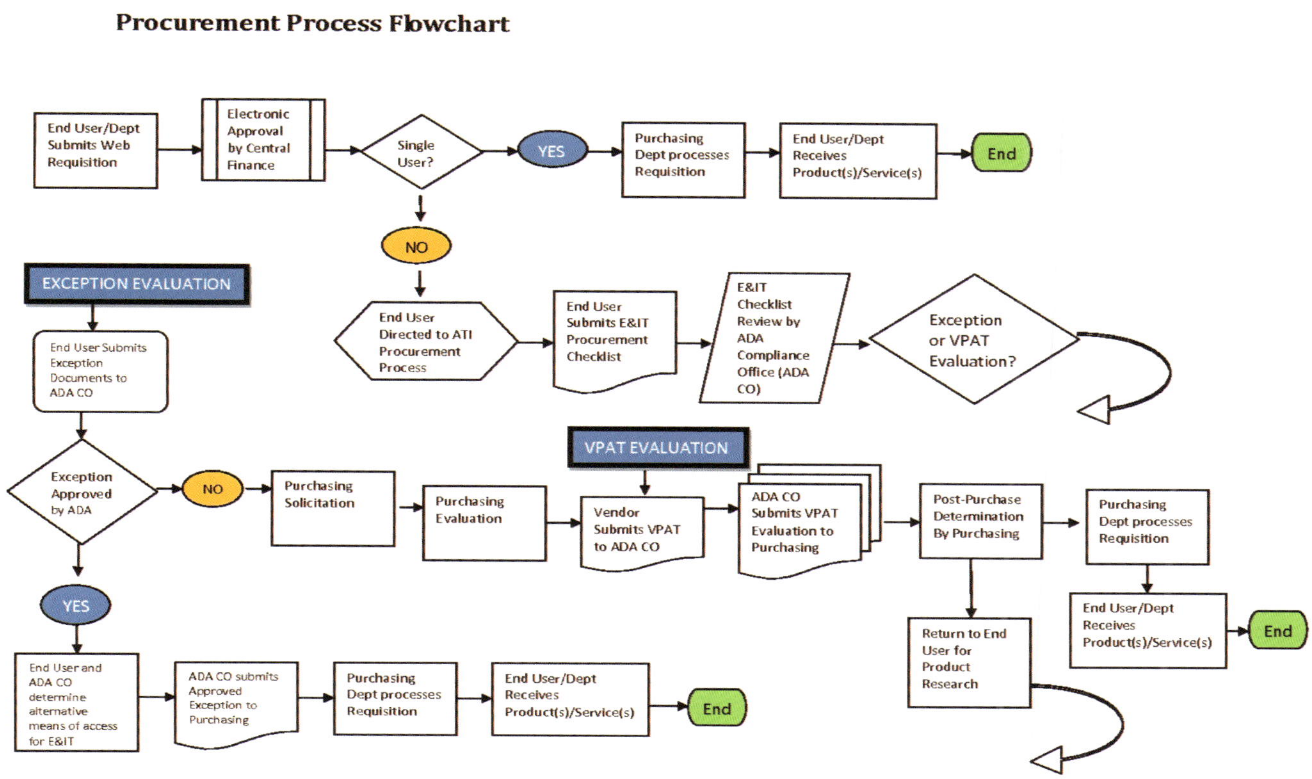

The above flow charts are typical and are used to document a process. A visual representation is worth a thousand words. By utilizing these charts, issues with the current processes can be immediately revealed and then you can start to use the 5 Whys to understand, why the outcomes are the way they are. This is very useful since when you proceed to the Future State you will be able to eliminate these issues and improve the throughput which reduces the costs of carrying out the process. In other words, eliminate waste-Muda.

Flowcharting can be done in Power Point, Excel and Word. If you prefer, you can buy for a reasonable cost, a flowcharting program or create your flowcharts with the applications as mentioned earlier in this chapter. Whichever you chose, having all of your processes defined will immensely help you and your team better understand the current state, the problems that are associated with the process being evaluated and what the future state should look and perform like. Flowcharting your future state incorporates the required activities to support you implementation.

Now that you have done the current state process via a flowchart, you now can develop the processes that you will be using in the new ERP system. Having this knowledge can cut down the time that is required to move the implementation process ahead and to also meet the implementation schedule's targeted milestones. Cross functional processes will have a wealth of pertinent data so that all of the team players are on the same page. When the outcomes are handed off to the downstream processes, there will be a higher likelihood that they will work with minimal rework and or customization.

As was done in the Current State flowcharting process, you will now look at how you can improve the current state and also look at the similar process that are incorporated in the new ERP system. The new process can greatly change the way you do your tasks and can streamline them as well. Above all, be open minded and willing to change! As it was stated, we are not implementing the new ERP system to replicate issues that the legacy system had. Job functions can change as well and most likely will. Since new process flows will be created, certain jobs could be eliminated. But there can be instances where there will be added processes and those who no longer do their current function can now be redeployed to perform the new tasks in the ERP system. It is by far a win, win scenario!

Once the Future State flowcharts are completed they can all be posted on the implementation room's walls so that you can now see how the process flow works and where potential problems may occur. It is very gratifying at this point to understand that you are now looking at how the "Enterprise" system will operate and how each process is dependent or independent on the other processes that are using the data within the database. A well laid out process evaluation process utilizing flowcharting will help ensure that the sum of the parts will add up to the whole.

CONCLUSION

The current state flowcharting is an exercise that outlines the way you currently operate with your company's legacy system. You should be able to see where the waste is Muda by asking the 5 Why's. This visual representation of the current state is worth a thousand words and helps streamline the entire process improvement task.

The Future State flowcharting creates the model that will be used within the new ERP system and will greatly improve the throughput by eliminating waste in the enterprise system. The old processes from your legacy system will be eliminated by the ones in the new ERP system. By all means, be open minded to Change and Continuous Process Improvement!

By viewing all the elements, the implementation team will now have a picture on how the database will be used and how the processes are going to function with each other. The processes will then be well synchronized, ultimately minimizing the need for reinventing the wheel.

NOTES:

CHAPTER 12
WHAT COULD POSSIBLY GO WRONG?

The following are some issues you need to look out for before, during and after an implementation. The format is to present the area of concern and the pitfalls associated with it. Then we will take each area in turn and present to you methods for mitigating these issues. The issues reviewed here are not meant to be all inclusive, but do represent major areas of concern that need to be addressed. Your list will be longer.

So you have done your homework and you have checked everything, but what can go wrong? All too often there are items-issues that pop up that weren't apparent to you during the implementation process. As we know, Murphy is always hanging around to spoil the show. So let's look at why it is important to do things right from the beginning!

You may ask why the concern? There is a cost for not doing it right the first time. There is a cost! The cost that we are talking about is the Cost of Quality.

Whether the failures are internal or external, an ERP system that has data processes, and work instruction errors will have a significant impact on the entire organization. There are costs that will be incurred due to these errors. Internal errors can be associated with routings, bills of material, master lists, time and attendance, and many other potential process errors. We must remember that the data that is shared, if in error, affects all downstream processes and operations.

All data analytics should be driven through the new ERP solution. Any attempts to work outside the ERP system should be discouraged!!

External costs are those that affect processes that are outside the four walls of your organization. These are supply chain, logistics, replenishment schedules, lead time, and others. If there are internal and external errors, this greatly compounds the ability of the organization in delivering a quality product and or service on time and within the budget. Appraisal costs are those where you have to analyze each of the errors that occur and then list them and ask why they occurred? This can be time consuming and takes away resources from the daily operations of your organization. This is definitely a Non-Value-Added process!

Last to be considered are the prevention costs. After performing the analysis of what happened, and why, it is then necessary to develop preventive actions to ensure that these errors do not repeat themselves going forward. This adds to more process procedures and lengthy training needs. So in the long run, errors must be kept to a minimum. Remember, errors that occur have an adverse effect on other areas in your company that uses the ERP system.

STOP

Now let's look at lessons learned here and what should be used to reduce these Non-Value-Added issues.

In a Lean environment, we use the Kaizen techniques to improve the process. The definition of these two terms are used interchangeably since they have the same outcomes and process improvement objectives. An ERP system is a collection of data, processes, rules, and work instructions. If any of these are inaccurate, then they need to be re-examined and improvements must be made to the processes, rules, and work instructions. This is an organizational issue and there needs to be those who demand that change takes place to ensure that the data, processes, rules, and work instructions are the best they can be!

By improving the processes on a continuous basis, errors that are eliminated in one area can eliminate many errors in many others. Remember, bad systems-processes beat up good employees. So it is in the best interest of all the end users to take Ownership and be Empowered in making the change happen! Remember, if you always do it the way you always have, you will always get what you always got! Don't accept less than ideal outcomes. So, the chapter's title is, 'What could possibly go wrong?'

BE AWARE

Let's look at a short list of items that can go wrong if the information-data is not correct.

BILL OF MATERIAL-BOM

If the data is incorrect—

- There could be wrong revision levels,
- Unit of measure
- Quantities required
- Product structure
- Specifications and materials
- Lead times
- Make or buy item

ROUTING FILES

If the data is incorrect there could be wrong machine centers or ones that do not exist, wrong run and setup standards, alternate machine center options and alternate routing options, listing of tooling needed for the operation, and inaccurate gauging needs.

SALES AND OPERATIONS PLANNING

If the data is incorrect there could be wrong sales forecasting or bookings, misleading sales planning, erroneous production planning, wrong rough cut capacity planning, and incorrect priority planning.

INVENTORY

Some concerns are questionable on-hand quantities, allocations, and failure to show inventory transactions in one or many warehouses. There are also stock adjustments and unexplained root causes for these adjustments. There could be other concerns.

The objective here is not to have the reader read a laundry list of items that are incorrect, but to make a strong statement that these data points need to be accurate!

If we look at the items mentioned above Bill of Materials-BOM, Routing Files, Sales and Operation Planning, and Inventory, all of this information-data is a shared resource for planning, accounting, engineering, manufacturing, purchasing, and shop floor management. Any errors incurred will have a direct effect on all of the ERP inputs and outputs that the end users have to perform their respective processes. Accuracy and integrity is of the utmost importance!

The question, what can possibly go wrong, can be quite a challenge unless you come to grips with why they occur. The Cost of Quality that includes internal and external failures, appraisal costs, and prevention costs can be quite expensive and slow down the entire process. We must understand the errors incurred will affect downstream processes and the errors will multiply and have a higher impact on the enterprise.

The use of Kaizen and Continuous Improvement Process techniques can improve the outputs of these processes, rules, and work instructions. This is everyone's business to continuously improve the process for better outcomes!

We will in the next section of this chapter discuss how we can develop solutions to eliminate these errors, in processes, rules, and work instructions.

We can see that there are a lot of components that make up an ERP system yet it is the data that is paramount. The accuracy is essential and you must analyze your current processes and ensure that they will meet the needs of the ERP system that you will be implementing.Using small pilot databases will help you understand the Current State of these databases and it will be much easier to correct than trying to migrate the entire database into the new ERP system. Simple is better!

The use of a Continuous Improvement-Kaizen regiment will ensure that you maintain and improve all of these issues during data migration, developing your SOPs, and measure the outputs of the ERP system you will be using going forward.

The following are some issues you need to look out for before, during and after an implementation. The format is to present the area of concern and the pitfalls associated with it and then we will take each area in turn and present to you methods for mitigating these issues. The issues reviewed here are not meant to be all inclusive, but do represent major areas of concern that need to be addressed. Your list will be longer.

SYSTEM PARAMETERS AND VARIABLES

System Parameters and Variables, what are they and how can they either help or hurt us? We don't think enough time is spent on this early phase of the implementation. This is early on in the implementation and generally occurs after the ERP software has been selected. In your ERP software package, this may go by different names, but their purpose is generally the same. Depending on the system you have selected, you can have a few dozen to a hundred or more of these variables or parameters. These variables are the building blocks of the system's output. It will provide the methodology for how the system will look at certain data sets. It is critical that all users understand how these variables were set for those outputs that they will be using in their department. Actually, it is critical that the users be involved in the decision making of how these variables be set up to begin with.

It is imperative that they be setup in a manner most recognizable by the users and if they can't be then they need to be test by the user to understand the output and how they will react to the output in the live database. By way of example, if you have the planning horizon set to look further out then you did before you may be buying or scheduling work orders long before you need then. The system will generate planned orders and if your staff normally creates actually order from planned orders it could lead to inflated demand that may be unreliable because it is out beyond your forecast capabilities. The order modifiers may be set to bring in too much material too soon. This would naturally lead to inflated inventory. The system variables and how they are managed are critical to the overall success of the system and the acceptance of the output by the end users. These variables are usually grouped by discipline or by the department that will mostly likely be the end users. The chaos that will ensue if you do not address these variables early on and get the right people involved with be formidable.

PRODUCT CHANGE MANAGEMENT

Well, like the old saying goes, the only constant is change and if you have ever worked in a manufacturing environment you are painfully aware of this fact. Whether your system is equipped with a change management module or not, managing change is critical, as it is a crucial input to the calculations of MRP within and ERP system. So you might as well embrace the horror now and find the best and most efficient way to deal with and manage change. If your ERP system is equipped with a change management module, it is critical that this module be tested and put through its paces. Make sure that you have tested its ability to handle change appropriately by using real world examples. Please ensure that all departments that could be a part of a change or affected by one change are included. There are many different types of changes most frequently generated by Engineering. These changes may be product improvements, safety or cost driven. More often than not, they will impact inventory, purchasing or the current build process or all of the above. The cost of a change should always be known and should be part of the implementation decision process. We are more concerned with how the change will be treated by the ERP system. An improperly executed change, will have a negative effect on the cost of the change, the timing of the change and may have safety liabilities associated with it.

Whether you have a change management module in your ERP system or you have created a procedure to manage a manual change off line, it is important that the change be managed with all the affected people involved. The system somehow is updated to manage the change and that all actionable parties are executing their portion of the change in a timely manner. Having meetings with all parties to kick off a change is very efficient. This will ensure that the actions of the responsible parties are executed in a timely manner. Any mishandling of a change can lead to negative outcomes including; improper disposal of inventory, safety and or liability issues either internal or external.

WORK ORDER MANAGEMENT

Work order management, or in this case the lack there of, can and does lead to many issues that can be devastating to the reliable output of the ERP system. Many departments rely on work order accuracy, including accounting. This is a key element to ensure they are properly managed using what is referred to as "Cradle to Grave." Tracking work order closure in a timely fashion is critical! If a work order is completed and the completed quantity is less than the schedule order quantity then you have a situation where the system thinks you are getting more of the item and in fact you are not. In this case, a planned order may not be generated when one is in fact needed. Once issued, the responsibility of the accuracy and ultimate timely closure of the work order falls solely on the Shop Floor. They must accurately report yield and material usage ensuring all operations are started and completed on time. Keep in mind that all uncompleted work orders are considered work in process (WIP). WIP is used by accounting when they are valuing inventory.

Work In Process (WIP) and Perpetual inventory are what make up total inventory. So in addition to sending the wrong signals to the ERP system now you are also inflating your inventory value. This is a big problem should you get audited or if you take your own annual inventory. Work orders are also analyzed by accounting and by Materials Management for material variances.

A material variance occurs when the amount of material issued to a work order is different than the amount it should have used, based on the standard created by engineering. This also impacts the planned usage rate of the material and can lead to a stock-out or over stock situations. Remember, that overproducing to a work order is just as bad as under producing. Both point to some error in yield, based on materials issued and need to be reviewed and corrected. Errors in standards on a work order, which is pulled from the routing information, will lead to inaccurate messages sent throughout the ERP system. These errors can be labor associated or material. Either way they will lead to poor and inaccurate planning going forward, which will undoubtedly be costly for the company not only in monetary terms, but in customer service terms as well. These will lead to poor overall company performance. If your company is a publicly-owned company these issues get magnified ten-fold. You will be under the scrutiny of an outside auditing entity and as such will be held to a higher standard. The direct costs with a poor audit are more obvious when you take 3 to 5 business days to complete and then are told you need to do it all over again because of major discrepancies!

MASTER PRODUCTION SCHEDULE

The Master Production Schedule (MPS) is often a major component of Sales and Operations Management. The purpose of the MPS is to translate the Sales forecast into a Production Plan that must be executed by the organization. The MPS is the demand side of the equation and as such, must represent the customers' needs. The MPS is a key driver in your Rough Cut Capacity Planning process. The sales forecast can be fraught with peaks and valleys that does not lend itself to a level loaded workflow though the organization. If these peaks and valleys are not addressed during the MPS planning process it will result in complete chaos on the shop floor and widely fluctuating cash flow for the company as expenditures will be tied to this plan.

The shop floor will experience periods of down time and then periods of extreme overtime where the schedule will not be accomplish due to constraints that fluctuate from being idle to overload conditions. This impacts both labor and machine capacity and makes for a very costly and ineffective operation. Any errors at the MPS level will be magnified as it trickles down through the entire planning process. Any minor error at the top level will translate into errors for Planning, Purchasing, Manufacturing, Sales and Customer Service.

The Master Schedule, when done properly, will enable the Master Scheduler to pull forward overloaded time frames into under loaded time periods. In this way the Master Scheduler can give manufacturing its best chance for success. This is known as being linear. Linearity needs to be a KPI and its achievement and level of achievement tells responsible managers how real their constraint/s may be and their duration.

With this knowledge, decisions can be made for the short and long term mitigation of these constraints.

CONSTRAINT MANAGEMENT

This issue plays into what we just discussed under the MPS portion of this segment, but it pertains to the issues surrounding improperly managing your constraints whether they be labor, machine or material. A constraint is what it sounds like. It is generally a lack of something that is keeping you from achieving a predefined schedule. The problems arising from not properly managing you constraints are many. The obvious is the inability to achieve any schedule. The less obvious is the inability to see trends and identify opportunities for growth. When you can't see these things, they also lead to missed opportunities! This will also have a negative impact on the big issue of on time delivery to the customer. If you can't be on time to your internal customer, you cannot possibly be on time to your external ones. This information can also be used to support capital purchases and support the need for more direct labor. Without proactively identifying these constraints, along with their respective severity and duration considerations we cannot hope to formulate a resolution to these issues. If you cannot quantify a problem you cannot solve the problem!

The following chapter offers solutions for what can go wrong. These are meant to suggest to you possible corrective actions that may work for you as is or with some modification. This will work to resolve your particular issue in your particular set of circumstances. We will take each subject in turn and propose a course of action.

CONCLUSION

To sum up, it should be obvious at this point, that there are a lot of variables that need to be addressed, which should be no surprise. We are dealing with the complexities of an entire system, one that we are attempting to not only run the company on, but run efficiently and effectively. This chapter should have given you a good taste of the overall process and many of the pitfalls to look out for. It is always a good idea to question everything and validate that the process works and produces an outcome that is conducive to a high quality productive system where data accuracy is king.

NOTES:

CHAPTER 13
SOLUTIONS: SYSTEM PARAMETERS AND VARIABLES

ERP SOFTWARE ENVIRONMENT

While the amount of system variables and/or parameters may appear to be a daunting task it is best to invoke the help of the new software provider and the entire steering committee. If you are working in an ERP environment and are upgrading be sure to see your IT department for a list of current variables and how they are set. That will give you a strong base to start from. The new software provider can best define the individual parameters and explain to the team how it will react in the system. It is best to break this down by department and work each issue to resolution in each function. You can have all departments work in parallel initially to expedite this process to its conclusion. You will be using your test database to test the output once you have made some informed decisions on how to set the variables. Be sure to review all options when looking at the variables. Remember to not just automatically set them to what is most similar to what you had before, if you had anything before, with a better understanding of how these variables work. Look for opportunities to improve. Challenge what you had before and validate that your selection is the best, most cost effective or efficient for the company, not just your department. Try to have a more global perspective when evaluating these options. It is important to realize that if an extra dollar spent in your department saves two dollars in another, then you're making the right business decision!

PRODUCT CHANGE MANAGEMENT

Within the Engineering module, a company has the ability to change and track product revisions. It is in this area where we need to have a written procedure and work instruction where appropriate. If your new system has a module to manage product changes, then it will be easy to flow chart and write procedures for. Engineering is usually the only ones that can generate a change via a form (hard) or electronic version (soft) most often referred to as an Engineering Change Notice (ECN). Other departments may request changes and it is best to have a form for this as well as for tracking purposes. This can be an Engineering Change Request Form (ECR). Make sure the form gets routed though all affected departments and that the form identifies all affected departments. Utilizing a workflow process in the ERP, if the system has it, for tracking ECN's. There should be a section to identify the priority of the change and the proposed cost supplied by engineering with input from any other departments as necessary. We won't get into the details of the form here, but you should get some input from the new software provider and they may have some sample forms with which you will you start.

BE AWARE A weekly ECN meeting may be necessary, especially if the company is new to the process so that all concerned parties understand their role in executing the ECN effectively in a timely fashion. Communication is key in the implementation of any change and this is a critical aspect of the follow through on this process. Change management will often be the responsibility of the Engineering Department to ensure the actions necessary are being taken. Often production control, purchasing, quality and materials will have key roles in this process. Remember, communication and follow through are imperative. Be sure all are well versed in the written procedure and understand their role in this process!

WORK ORDER MANAGEMENT

Work order management and its impact on the ERP system is sometimes overlooked or even trivialized, but make no mistake it can render any ERP or MRP system useless when it is not properly managed. The cradle to grave management is performed by the shop floor and closely monitored by Planning and the Accounting department.

You must set up KPI's to keep this process on track. One KPI to use is on time completion of the work order. You will be comparing the scheduled completion date to the actual completion date. Shop floor management needs to track percent on time complete and post their performance on a monthly basis. The shop floor personnel such as supervision or leads need to track this on a daily basis so that any deviation can be quickly addressed and resolved. All missed dates, and it is ok to give yourself a small grace period here, especially at the beginning, must be analyzed and explained. A root cause should be identified and a corrective action put in place. You need to make this a meaningful part of shop floor control and part of their everyday job functions.

IDEAS We have seen it work best when it is made part of their performance review, inclusive of shop floor personnel and shop floor management. Material and labor variances must also be tracked, analyzed and any significant deviations to the standard must be identified through root cause analysis and a corrective action put in place. These two activities, more than anything else, will put you on the right track and your data output will be accurate and timely.

MASTER PRODUCTION SCHEDULE

There are several ways to analyze your level of performance with the master schedule and how it is translated into an effective production plan. The MPS must be in support of the customer demand and can only move if the customer demand moves, for no other reason!

The question is how linear is your production plan? Achieving linearity can be done but is not often easy. It takes a good understanding of your product/s and their labor and material content. If you are making widgets and each widget was made up of the same material and took the same amount of time to make, then you can check and measure your linearity in number of units. Remember to support your customers' requirement. Your shippable product must be done on or before the required ship date to be on time to the customer.

Linearity is akin to level loading. In a more complicated or diverse product offering, the overall load a product represents in hours may be the way you need to track linearity. Depending on the product mix, you may have 20 units in one week and 50 units in another and yet they both represent the same load to the shop floor. In this case you are one hundred percent linear for those two weeks. Linearity ensures that a consistent load is presented to the shop floor affording them the opportunity for the greatest chance for success in achieving on time completions and on time delivery to the customer. Managing and reporting on your linearity performance is the single best performance measurement for managing an effective MPS.

CONSTRAINT MANAGEMENT

One point we cannot stress enough, is you never schedule your MPS, or your finished goods items, if you don't have an MPS, based on your constraints whether they be labor, material, or machine. You always schedule to support your demand and level load as best you can and then you manage the supply side to meet your demand whether your constraint is labor, material or machine. To do otherwise, is to grossly mismanage your business and to hide issues that need to be addressed that could improve or grow your operation and your business. Productivity is the age old tool needed to manage your constraint. You need to track Labor and Machine productivity!

The formula for Productivity = Efficiency x Utilization (P = E x U) to identify your productivity coming off your equipment. Labor efficiency tracks Actual to Standard on work orders and that data should be captured on a daily basis. Once again, there needs to be a process for identifying significant deviations that need to be explained, while Root Cause analysis is done and a corrective action put into place. Corrective action and preventative action needs to be considered.

BILL OF MATERIAL-BOM

The BOM is a critical and definitive listing of all of the information about the products that are going to be produced by your company. The BOM is then used by the Manufacturing Resource Planning System-MRP to provide you with what needs to be ordered and when based on lead times and build rates. Essentially, we say that the "BOM" is exploded through the MRP to its lowest levels and the MRP regenerates all of the data within the system to give you a "Net" report of what is needed in order to make the build rate per the customer's demand rate. If the BOM is not accurate, it will give the MRP false signals and the planner will order incorrectly and will not meet the demand rate.

Since you will be using your legacy system's BOMs for all of the products you manufacture, how do you ensure that they are correct? You should have, through experience, noted that there are issues with the "Current State" and they should all be noted so that when you begin your implementation process they can be brought up and resolved. This is very important. The members of the team responsible for this portion of the implementation should be well versed subject matter experts-SMEs, to know that these issues exist.

At this point, we did not want to take on more than we can handle. Take a small sampling of different "Product Families" and migrate the BOM database into an Excel spreadsheet that can be used to populate the ERP system. By doing this in smaller bites, the obvious will be seen and you will probably see the errors in the BOMs a lot quicker. Here is where you need to make the necessary changes so that when you bring over more of the BOM database, you will have a higher level of confidence that it is accurate and usable.

When you pilot the planning and purchasing functions within the ERP system, you will see if there are any other issues that may exist as the result of the MRP regeneration. This then can help you formulate a plan of action on how to mitigate these issues and keep the file size manageable. As you refine the BOMs, the MRP will give you a more accurate picture of what needs to be done to meet the customer's needs.

The following issues should be taken into account when you review a BOM:

- Levels of the BOM
- Material required for components
- Is it a Make or Buy item or can it be both for flexibility needs?
- Items per level (wagon needs 4 wheel, 1 axle, 1 handle, 1 body, etc.)
- Lead times to buy or make
- Specification requirements (Mil, J, ASTM, etc.)
- Planner responsible for the BOM-Program
- Item Value: A, B, or C

MISCELLANEOUS ITEMS-DATA

If the BOM is accurate, then there is a high probability that the correct ordering requirements will be obtained from the MRP and the correct ordering actions will be taken.

SALES AND OPERATION PLANNING-S&OP

What does this all mean? In very simple terms, it is what and how the company is going to meet its commitments to their customer demands. In addition, it also takes into consideration future plans to expand and or take on new product lines. Last, it is the balance of the Demand versus the Capacity.

As result of this analysis, demand is defined. This is used to generate a MPS that will drive the MRP. If this is a broken process then the results from the MRP will be inaccurate, costly, and you will not meet your customer's demands! Management must review these issues with the company's S&OP and right the ship. Remember, this is an enterprise system that is driven by demands from the customer. If the demands are incorrect, then how can you plan the capacity, the materials required, lead times, and other requirements needed? This must be addressed first and foremost!

ROUTING FILES

Routing files are a result of the way a product-component is going to be manufactured. If it is a buy item then it is quite simple, there are no internal routings to define. Those components that are a "Make" item then need a specific routing file to ensure that they are manufactured correctly and within the lead time required to meet the current customer demands. In addition, this will determine what capital equipment is needed in order to utilize the routing files created.

In addition, to this, there may be a need to have alternate machine centers and or alternate routing files. Why? In some cases there may be a need to utilize an alternate routing file or machine center to mitigate bottlenecks in the manufacturing facility. If you do not have, or have routing files that are not robust, then you will be limiting yourself to what you can do. In addition, the MRP scheduler will produce less than adequate dispatch reports and inhibit your ability to flow product throughout the manufacturing facility as well as outside facilities to produce your products on time.

? *Review all of these files and where possible make the necessary improvements including changes to meet the demands you will be required to satisfy.*

INVENTORY

An "A" class MRP requires that the BOMs be 99% accurate, Routing Files be 99% accurate, and the inventory be 99% accurate! Why? The more accurate these are the better the chances are you will be able to meet your customer's demands and the outputs from the MRP will be timely.

Inventory is both an asset and a liability. Of course, inventory has value and you must have the right amount of it on hand for the monthly plan. On the other hand, it is a liability if you have too much on hand and it takes away capital that could have been used for other investments to meet the S&OP. Inventory accuracy is then the name of the game here. If your inventory accuracy is inaccurate, the signals from the inventory to the MRP will be wrong and thus timed orders will not be correct to meet the demand needed.

Inventory also must be classified by its value. Inventory can be classified as A items, B Items, and C Items. A items are the most costly and usually have the longest lead time to produce and to replace. B items are less expensive, but can be an issue in meeting the demand. C items are the least expensive and are the easiest to replace as compared to A and B items. Thus the BOM must have each of the items defined by an A, B or C classification.

Once your active inventory has been defined and you have the unit cost identified the next step is to resort the group in descending dollars by unit cost. At the bottom of the unit cost column sum the entire column in dollars. This sum is not the total value of your inventory that can only be determined by summing the extended cost column, which takes into account on hand quantities. For this exercise we are just looking at the sum of the unit costs. It is with this sum you will use the 80, 15 and 5 percent principle to determine A, B and C code classifications. Multiply the total sum by 80%, 15% and 5% this will give you your dollar thresholds for the next step and that is identifying parts for each A, B and C code. With your spreadsheet arranged in descending dollars, highlight from the top down in the unit cost column until you have reach your first threshold, which is the dollar value representing by your 80% calculation you determined earlier. From the point where you stopped to the top of the list is your list of "A" items. From the point you left off at the end of your "A" list begin with the next items and highlight the column until you reach the "B" item threshold determined by your 15% calculation. Follow the same process for you "C" items. This should work out to approximately the following:

A items comprise 20% of your inventory but are 80% of the inventory value. B items comprise 65% of the inventory and are 15% of the value and C items comprise 15% of the inventory and are 5% of its value. So as you can see, classification is required to understand the makeup of your inventory.

Inventory is also classified by Work In Process-WIP material and or components in the manufacturing facility, assembly and test areas. Inventory on hand is just that. How much can we draw on to meet the demand? If the inventory is not accurate, you may find "Stock-Outs" and you will not be able to meet the demand. You will need to expedite the next lot or lots to meet the demand. This adds cost and reduces the margins expected.

To ensure that the inventory is accurate you should perform "Cycle Counting" by an ABC analysis (A items, B Items, and C Items).By doing so you select so many A, B and C items in your inventory and count them on a cyclical basis. In doing so you will identify where the issues are and how many times they occur. Then corrective actions as well as preventative actions can be taken to minimize the inaccuracies in your on-hand inventory and your WIP. This is standard inventory practice and it really works! If your inventory is accurate, your MRP outputs will also be accurate and the correct ordering actions will be taken to meet the demand.

CONCLUSION

In conclusion, we would like to say that we can go point counterpoint all day on the vast amount of possible scenarios that could occur during an ERP implementation. There are at least as many things that can go wrong as there are software packages to choose from. Hopefully, some of the issues we covered will serve as fertile soil for you to think about you own unique set of circumstances that will require their own unique solutions. We not only attempted here to present you with real world issues and their solutions, but a foundation on which to build your own critical thinking and problem resolution skills. Remember, you are not alone in this. Seek the input of your peers, the new software reseller's expertise and groups and organizations that no doubt will have someone that dealt with a problem similar to yours, if not the same as, the one you are facing. Use every resource at your disposal and you will persevere and succeed.

NOTES:

CHAPTER 14
SUMMARY AND PATH OF CONTINUOUS IMPROVEMENT OUTLINE

1. IMPLEMENTATION OF AN ERP SOLUTION IS AN EVOLUTION AND NOT A REVOLUTION

- Understanding issues with the current legacy system
- Not an enterprise system
- Database is not a relational database
- Manual reports taken from the legacy system to run the enterprise
- Does not support your global supply chain needs of your customers
- There are no more upgrades
- Cannot communicate with your customer's ERP system

WEAK IN PROJECT MANAGEMENT SKILLS AND LEADERSHIP

Whether you are replacing your legacy system or an existing ERP system, implementation of a new ERP system is critical to your company's future in this very competitive global marketplace. Even though, you have determined that a new ERP system needs to be implemented, you must understand why the change is needed and what not to duplicate when you launch the new ERP system. A checklist of issues that you currently have with your legacy system should be drafted so that you can understand its short comings and what you will need to look for when you solicit ERP solution providers and ERP resellers. This will be critical, that this checklist be representative of the end users' needs, how you will support your internal as well as your external customers, and introduce a Kaizen approach (Process Improvements) to your standard operating procedures-SOPs.

2. WHAT SHOULD BE THE DESIRED RESULTS?

- A true relational database
- A system that is inclusive of all of the tactical groups in the enterprise
- Accurate data using the reporting that is available within the new ERP system
- Is expandable and upgradeable
- Can interface with your customers and suppliers ERP solution
- Establish a workable and obtainable plan within the budget guidelines agreed to at the start of the ERP implementation.

After you have drafted a "Checklist" of all of the issues and desires that you will be looking for in the new ERP system, the new system that you purchase should be seamless and satisfy all of the end-users needs to do their jobs more efficiently while satisfying your ultimate customer's needs. The ultimate goal is to have a true relational database and that it is ACCURATE! An ERP system is just what is says. It is an enterprise system. Pay close attention to this pointer! The goal is to establish an enterprise system that will allow your company to meet your customer's needs and allow it to become more competitive in today's global marketplace. Accurate and timely data is an essential requirement in order to be an organization that has the ability to be an "Order Winner".

3.WHAT ARE THE BARRIERS WITH THE CURRENT STATE?

- People
- Lack of IT expertise
- Lack of leadership
- Legacy databases is not whole and accurate
- Everyone is on another page as to the state of the company
- Training
- Language
- Knowing how to use a Windows OS
- The cost to upgrade the user's PCs to the new OS that is required for the ERP Team Structure.

One of the biggest barriers in an ERP implementation process is trained and skilled people. After having used your current legacy system for many years, there is a tendency not to think outside the box! You will need to have the project's sponsor, champion, and project manager be the "Change Agents"! They are the ones who will be driving the bus so that all who are on board understands their responsibilities and the ERP Project Charter's requirements, ensuring that the project will have a successful conclusion.

Training is a critical issue as well. You can have the ultimate ERP solution, but it is the end-user who has to understand how to use what the system is telling them! Remember, an ERP solution is not a silver bullet! It is a collection of data that informs you what the current state is. You will then mitigate the issues identified by those who have the skill sets to introduce solutions and can follow-up on the outcome of the solutions. Training should have a budget within the scope of the ERP implementation project and employee training must be included. Skimping on the training budget and time allotted for the training will lead you down a path to failure. This is a team environment and everyone needs to be qualified in order to reap all of the benefits of your newly implemented ERP solution.

4. SELECTING THE ERP SOLUTION AND ERP RESELLER

When you start looking for an ERP Solution and Reseller, this can be a daunting task. Yet this does not have to be if you do it in an organized fashion. If we refer to table 6.1, ERP SOFTWARE COMPARISON, you will see there are major characteristics that you should be looking for when you compare the various

ERP solution providers and resellers. It is important to filter your solicitations to those solution providers who are familiar with your industry! There are many choices and each solution provider will be more capable than the other in providing your company with the right ERP solution.

IDEAS *Remember the budget too!* The purchase and implementation of an ERP solution is expensive and you should have a budget benchmark when you start to solicit an ERP solution provider and reseller. You may not have the funds or the staff to implement all of the modules that you want. So implement the modules you need to run the business and support your customers first. Then on a cyclical basis add in those modules that will enhance the core system that you have implemented. In this instance, patience is a key trait that you will need to develop to ensure that you achieve your expectations! Patience is a virtue!

5. DEVELOPING THE LEADERSHIP TEAM FOR THE IMPLEMENTATION PLAN

- Who are the Sponsor and Champion?
- Sponsor has to be the CEO or similar in responsibility
- Champion is selected by the sponsor with explicit guidelines

The champion and the sponsor are the ultimate persons held accountable for the success or failure of the project.

As it was stated in this guide, leadership is of the utmost importance. In our Lean teachings we say "People need leadership and not management"! [6] What this is stating is there needs to be a leader at the helm of the ERP implementation project! This is where the buck stops when critical decisions need to be made. This is the Sponsors' responsibility. Second, the Champions' responsibility is to carry out the mission-charter of the ERP implementation project. The champion is the eyes and ears of the sponsor and it is his or her responsibility to ensure that all barriers are dealt with and are resolved. If they cannot be resolved, then these outstanding issues need to be brought to the sponsor for resolution. These issues-barriers must be dealt with as soon as possible so that they do not have a negative impact on the project's timeline and budget!

- Who, and from where, do you select the Project Leader and Team Members from?
- The Project leader should be chosen by the champion based on his or her knowledge of the organization, its employees, and what the current state is with the legacy system being used
- Team members should be SME's in each of the tactical groups within the enterprise but others can be included to balance the view of the current state
- These individuals should have good team skills and must have a good attendance record and are willing to take on responsibilities outlined in the ERP project charter

PROBLEM SOLVERS AND TEAM PLAYERS

The selection of the ERP team's project manager and team members is crucial! The project manager must be familiar with how your company operates, what the issues are with the current enterprise legacy system, and be a Change Agent. The project manager's responsibility is to carry out the project's charter and to ensure that the project is meeting all of the milestones of the project's timeline. The project manager has to be a "People Person" so that he can negotiate issues that arise between team members. There needs to be a high level of agreement on how the project is going to proceed. This will ensure that everyone on the bus knows where they are going, what position they are responsible for and how they will get there. Teamwork is of the utmost importance!

Team members should be subject matter experts-SMEs in their areas of responsibility. Nevertheless, as it was mentioned in this guide, it is essential to have individuals who are not SME's to contribute fresh ideas from another perspective. Sometimes looking at the current state from the outside can bring to the table process improvements that were not seen by those inside the four walls of your company! By all means, the project manager and team players must be problem solvers. They were chosen because they have the skill sets and knowledge on how your company operates. They then are you best assets! Use them wisely!

- The Charter: Mission, Objectives, Deliverables Scope Benefits, Schedule
- A charter must be developed and utilized to define the entire ERP Project
- Each of the components of the charter are key requirements and or KPIs that will determined the outcome of the ERP project
- The charter should have KPIs defined that are measurable

DEFINE THE "WHAT" AND "HOW" OF THE PROJECT

We now have all of the essential elements of the WHO will do WHAT as it relates to the ERP implementation project. Now we have to define the What and How! This is done through the development and use of a Charter! The charter will be very specific and will define the mission, objectives, scope, deliverables, benefits, measurable, and project timeline. All of these have to be clear, correct, and concise! This is what will be followed throughout the entire ERP implementation project. The charter must be agreed to by the sponsor, champion, and the project manager. This document will be what management uses to monitor the project's progress. Key Performance Indicators-KPI's must be established early on so that they are measurable. What gets measured gets fixed! It is essential that KPI's be developed and monitored from the start of the ERP implementation project. With everyone on board knowing what is expected, the chances of having a successful outcome will be greatly enhanced.

- How do the ERP solution and or Reseller fit into your project team's needs?
- Does it close the "Gaps" that are identified in your legacy system
- Does it give your enterprise the competitive edge that you need to compete in a global supply chain?
- Does the solution provider and reseller support the goals and objectives of the ERP initiative?

Can they resolve or suggest solutions to issues that are causing the project to come to a stop?

Your company now has a charter, a project organization structure, and a need to implement a new ERP system. But what ERP solution and who will be the ones to assist you in the implementation process? These questions are critical and must be answered! As we discussed earlier, you will need to establish a checklist of the issues you are currently having with your legacy system. These issues that are identified on this checklist are the "Gaps" you currently have and want to close by implementing a new ERP solution!

The selection of an ERP system will depend on if it meets your company's needs as it relates to the industry that you are operating in; hospitality, manufacturing, retail, etc.! When soliciting ERP providers, what industry sectors are their strengths? When interviewing an ERP solution provider, ask them for references. Don't just accept just one or two. Try to get at least 4 to 5 end-users to get fact based feedback on the success of their implementation project and how the ERP system met their needs!

If there is going to be a need to have an ERP reseller involved, the same questions should be asked as mentioned above for the ERP solution provider? This is crucial since they will be an extension and integral part of your in-house ERP implementation project team! They will be the ones who will provide you with expertise and services that you may not have currently in your organization. A good example of this is database migration from your legacy system to the new ERP solution. Your team needs to understand and know what its deficiencies are so you can select the best ERP solution and ERP reseller for your project!

By all means, DO NOT JUST GO ON PRICE!! There are times when going cheap can in the long run will be very expensive. Our advice to you, is to be frugal and spend your project's dollars wisely, identify where you get the most value for your budget dollars. If necessary, review your project's budget and make the necessary changes when required. You may find that you might have to reapportion your budget dollars to where they are most needed.

6. ASSESSING THE CURRENT STATE AND CAPABILITIES OF YOUR EMPLOYEES

Do you fully understand what the Current State is and what needs to be done to improve the outcomes desired with the new ERP solution?

How prepared are your employees with basic materials management practices and are they prepared to make the necessary changes that will be required when they start to use the new ERP system?

One point that cannot be stressed enough and is a common theme in any successful implementation is "Training." Preparing your facilitators and your work force for an implementation is one of the keys to ensuring a successful implementation. Make sure you have done a thorough workforce evaluation and defined the training gap analysis, which will bridge the gap between their current knowledge base and where they need to be to take the organization to the next level.

7. CURRENT STATE, DATA INTEGRITY, AND CREATING THE PROJECT PLAN

- Have a full understanding of the current state
- Ensure that the database is valid before migrating it into the new ERP system

Have a valid project plan developed and ensure it has a high level of approval by the sponsor, champion, ERP reseller and project team leader.

When mapping different paths that need to be followed by each department it is essential you know you're starting point. The foundation of any data migration is the accuracy and the validity of the current data you are looking to load into your new ERP system. Garbage in Garbage out (GIGO).

DO NOT BOTHER implementing a new ERP system if you are going to load it with bad data from the legacy system.'

ERP is just a tool it can't make you do things correctly that is on you! Upper management support is very important to the success of any major project and is especially critical here. Upper management must have a fundamental understanding of the work that is entailed with an ERP implementation and must provide the leadership and commitment to stay the course.

8. CONFERENCE ROOM PILOT AND MEETINGS

  - o Have you mapped out a schedule of regular team meetings?
  - o Crucial to the implementation is an organized plan of meetings
  - o Meetings should be focused and mapped according to milestones in the project implementation
- A conference room pilot should include actual activities replicating what will occur when the system goes live.
  - o Do you have periodic management progress meetings scheduled to close the loop?
- Alleviate the guessing and put into the plans coordinated functional managerial area meetings.
  - o Are meeting minutes taken to record all of the actions outstanding, completed and issues that exist?
- Minutes are a necessity for accountability and directions required.

What are the conference room pilot process steps-procedures?

Remember the conference room pilot is the closest replication you have to being live. It is critical that the Pilot Program be used to exercise all the activities you anticipate experiencing when you go live. The most effective way to do this is to take a sampling of each of your products that will represents different process flows, as they currently do in your current business environment. You need to run these through the Pilot Program separately so that the process can be followed and assessed each step of the way without the interference or added complexities of running multiple products. This will allow for a cleaner and clearer process evaluation. Run each of the sample products in turn and follow to their satisfactory completion, fixing any process issues or concerns along the way. You need to document any obstacles and their resolutions.

9. FLOWCHARTING THE CURRENT AND FUTURE STATE AND WHY IT IS IMPORTANT?

- The current state and what does the analysis tell us?
- What is desired as the future state with the incorporation of the new ERP system?
- What needs to be change to ensure that the outcomes meet the desires of the future state?
- Is it possible and what will need to be done beyond the capabilities of the new ERP system?

The analysis of the Current State is crucial in defining the path to the future state. At this stage, you will inevitably be fixing existing processes in preparation to go forward to the Future State. Establish your goals for the future state. You need to have well defined and measurable KPI's. Target percentages need to be identified and published to the company at large on an established and adhered to periodic basis. Course adjustments will need to be made as you shake out the processes throughout the organization and to foster a culture of continuous improvement.

10. WHAT CAN WRONG AND HOW DO YOU MITIGATE THESE ISSUES AND DEVELOP A WAY TO RESOLVE THEM.

- Identification of issues is very important and they should be listed in a parking lot so that they can be reviewed and a plan of action can be developed to mitigate them.
- Implement the solutions, test them, and review the outcomes. Did they eliminate the issues?

LESSONS LEARNED

During the ERP implementation process, many issues come up and they all need to be addressed for resolution. Unfortunately, only so much can be discussed and resolved at any one time. You need to be aware to not allow Scope Creep and remain focused on the project's timeline and priorities. If this does happen, then you will be wasting precious man-hours-capacity and jeopardize meeting the project's timeline.

Root Cause and Corrective Action (RCCA). Be sure that all necessary personal are trained in how to identify root cause problems and how to implement solutions. It is critical that this process is done effectively and that we do not apply band aids to what will prove to be a much deeper root cause. This is not as easy as you may think and wasting time on band aids is just that a waste of time and will lead to addressing the same issues over and over again.

To avoid this, if many issues and or ideas come to the table when you are meeting to discuss issues with the ERP project, put those that are not immediately relevant in a "Parking Lot". This is the same as a place holder or bookmark that will be a repository of these items that were brought up at prior meetings. When the issue comes up at a later meeting you can review this parking lot and pull those items that are relevant to the discussion at that time. This will guarantee that the ideas are not lost and that they will be used constructively in a timely manner!

In this guide, it was stated many times that you must test every process to ensure that it works each and every time! Not doing so will put the implementation process in jeopardy and have a direct effect on downstream processes within the ERP system. Remember, if it doesn't work consistently, fix it! It will save you a lot of grief and time which can be spent in value-added work!

While you are going through the ERP implementation, you should take copious notes of what worked and what didn't. In addition, what did you learn from these experiences? Those copious notes will be an asset to the team and company. Often these experiences are not captured and the loss of this data is a major detriment to the implementation. By being cognizant of these lessons, they can be a very valuable tool to use in the future. Learning is a lifelong event and the lessons learned during the ERP implementation project is part of it.

11. BEYOND THE IMPLEMENTATION PROCESS: GOING LIVE AND CONTINUOUS IMPROVEMENT.

  - How are you going to monitor the processes of the new ERP system once you go live?
  - Maintain the project team for at least 1 month after the go-live date
  - Maintain an open line of communication with your ERP solution provider-reseller
  - Develop a system of checklists to monitor abnormalities
- Have system meetings on a regular basis to discuss issues that arise.
- Will you have a core team available to mitigate the issues you realize after you go live?
- Continuous improvement-Kaizen is essential to ensure that the ERP system is maintained and continues to perform to your expectations.

Working for months and then going live with your new ERP solution is a major accomplishment. It is very satisfying to see the entire organization work towards the mission in the charter you created many months ago. But it doesn't stop here. There is still some very important work that needs to be done!

With any new system, there are unexpected issues that will rear their heads when you go live. Even though you have piloted the system and proved out all of the standard operating procedures-SOPs, there is a good chance that something may not have been found and is problematic. So what do you do?

IDEAS The original team that worked so diligently during the course of the implementation process is going to be your safety net! They are the ones who are the most knowledgeable and can best deal with the issues that you encounter. Remember, whether you are using your legacy system or have just gone live with a new ERP solution, there is a business to run. Customers still want their products and or services on time and issues with your ERP solution must be dealt with expeditiously! Thus the internal implementation team and the ERP solution provider-reseller are the subject matter experts-SMEs. They are the ones who can see beyond the issues and resolve them quickly! This internal-external team should not disband and should be available for the next month after you go live!

You should also maintain an open line of collaboration with you ERP solution provider-reseller. They are the ones who have the most knowledge about the ERP solution you implemented and they are very familiar with your company's operations. SME's are the ultimate resources to get you through the early phases when you Go Live!

If needed, setup checklists via an email format so that the issues reported can be sent to a specific resource within or outside your organization. The collection of these reports will enable you to have a record of all of the issues that were experienced and you can see how each problem was resolved. This is a library of solutions!

In addition, to what was just stated, have regular team meetings so that issues that are experienced can be discussed. Meetings of this nature can be used to brainstorm the issues being experienced and where solutions can be found. Working in a team environment is advantageous and is very rewarding!

- Installing software upgrades and why is this important?
- To keep your ERP solution up to date you must regularly update your version of the ERP software
- If you have custom scripts-macros you will need to know if they will work with the upgrade
- Installing the latest revisions will ensure that your ERP solution will be current and will provide you with the most capabilities

To ensure that your ERP solution software works with a change of a Windows OS, you will need to make sure you have the latest ERP solution revisions installed on your server.

Software like hardware has to be upgraded to keep them up to date to the latest revisions and capabilities. When you negotiated the contract with the ERP solution provider-reseller there should have been a discussion on software maintenance. This is usually based on a percentage of the software cost. It is imperative that you maintain the revision level of the ERP solution's software!

If you needed to have custom scripts and or macros built for your company's requirements, it should work seamlessly when you go live. Unfortunately, when you upgrade to a new software version, these custom scripts and macros may not work as they did with the previous revision level. Make sure you review this with your ERP Solution provider-reseller so that you know how much it will cost and what are the chances that they may or may not work properly with the upgraded software revision!

Initially, your motivation to install and implement a modern ERP solution was to modernize your enterprise system! Unfortunately, what was good now may not be good tomorrow. That is why you have to upgrade the ERP solution's software on a continuous basis. PCs also have the same issues. If an operating system-OS is upgraded, the ERP software will also have a revision to deal with these changes in the end-users PC. In addition, if you need to communicate through customer portals, their system may not communicate with your ERP system because the software's revision level is out of date.

If you do not upgrade your software and you allow two or three versions to pass you by, you may not be able to upgrade because there is such a big delta between what you are using and what is the latest revision.

This can jeopardize the ERP system and you may have to do a complete implementation again. So the rule is, upgrade and do not leave yourself open to a scenario that will put your company in jeopardy! A few cents saved today can cost you many dollars tomorrow. Be frugal and look at the upgrade as an insurance policy. This will guarantee that the ERP system will always be up and running to service your customers! That is why you implemented the ERP solution in the first place!

WHAT EXACTLY IS ERP FAILURE?

- Implementation outcomes are not black and white but are evaluated on a subjective basis, so it's difficult to define failure within the context of ERP. Factors that contribute to an unsuccessful project include:
- Significantly exceeding the estimated budget;
- Exceeding the estimated timeline;
- Failure to achieve the projected return on investment;
- Failure of the system to fulfill objectives or deliver benefits
- Failure of the ERP solution provider or reseller to meet commitments.
- The miscalculation of your needs and or available resources
- The lack of employee training
- The lack of support from the top down.
- The inability to transfer legacy data to new ERP

If your project meets one or more of the above criteria—with the exception of the fifth example—it doesn't necessarily mean that you should scrap your ERP system or regard it as a loss. Many companies approach ERP initiatives with unrealistic expectations or underestimate how long it will take to see benefits or realize profits. In other words, what an organization perceives as a failure may not actually be a failure. [6]

IDEAS ADDITIONAL RESOURCES: A HELP DESK

Figure. 14.1 Help Desk

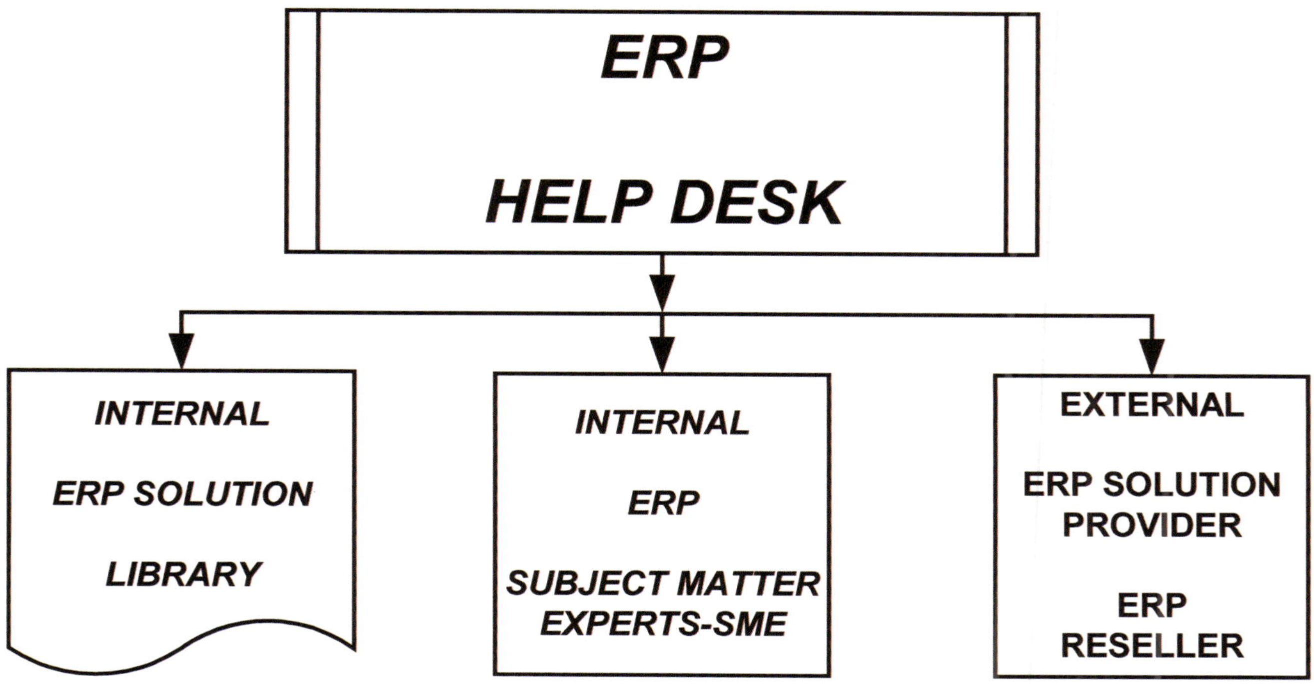

SOURCE

1. After you "Go Live", there may be instances where an end user may need assistance with an issue he or she may have while utilizing the ERP solution. It can be very frustrating in trying to navigate these! To help in these situations, we recommend that you setup the following "Help Desk" within your company:

2. Internal ERP Solution Library: A repository of documentation that has been collected during the implementation process should be stored in a virtual library. This includes policies, procedures, work instructions, and other documentation. This is secondary data and can be very useful in mitigating issues that you may have!

3. Internal ERP Subject Matter Experts-SMEs: In addition to your Internal ERP Solution Library, you can also put together an internal team of SMEs. Setup a phone directory within your organization with the names of those employees who are the ERP solution experts in each of the functional areas. This is a wealth of resources and they can be reached readily to resolve issues that arise.

External ERP Solution Provider and ERP Reseller: We also recommend that you maintain a strong relationship with both of these external resources to help you mitigate issues that are beyond the scope- capabilities of your internal Help Desk resources.

All three of these resources will ensure that you have the help you will need on demand.

CONCLUSION

Throughout this guide, the authors have been utilizing their past experiences to make you aware of the important issues in the preparation, section and implementation of an ERP system. The reader should have gotten a few pointers that resonate over and over again and should not be taken lightly!

First and foremost their should be a well defined charter that will guide the entire organization during and after the preparation, selection, and implementation of an ERP system is completed. Second, teamwork is an essential part to a successful implementation process. Everyone should be sitting in the right seat on the bus. Leadership should be hands on and they should ensure that they have allocated the necessary resources to ensure the project will have a successful outcome.

The selection of the right ERP solution provider and reseller is very important and they are your outside team players that will assist you in your endeavors. Make sure you have solicited a number of ERP software solution providers and resellers so that you can be assured that you have a bank of providers from which to select.

The authors have over and over again stated that data is king in an ERP system. Great care should be taken to ensure the data that is migrated from your legacy system is useful in the newly implemented ERP system. Remember, Garbage In, Garbage Out-GIGO. Also remember the Cost of Quality and what it will do if you get in wrong. A well devised migration plan with checks and balances will ensure that you get what you are expecting. If it didn't work in your legacy system don't replicate it in the new ERP system!

Education and training of you people is of the utmost importance! Do not short change this effort in either your budget or your timeline. The endusers of the new ERP system are the ones that will be interpreting what the system is telling them and they are the ones who will have to make decisions on how to navigate the Current State. An ERP system is not a "Silver Bullet". It is a means by the way employees make better decisions to improve the throughput and needs of your customers. Make sure your employees are the best they can be. They are the ones who will either make or break the successful use of your newly implemented ERP solution.

Understanding your current state using your legacy system is imperative. Why? Because you do not want to replicate the under performing or inaccurate characteristics in your newly acquired ERP solution. You must review all of the processes and their respective standard operating procedures-SOPs and determine if they will work going forward. If they don't, do not bring them along for the ride in the new ERP solution. As you work in each module, rewrite all of your SOPs so that they are reflective on the new ERP solution. Those processes that don't work, get rid of! Replace them with ones that do work and see how the new ERP system can streamline the processes that you will be utilizing to eliminate waste and improve the company's performance as it relates to meeting your ultimate customer's needs and wants. By all means, experiment and reflect on the outcomes before any SOP is set in stone!

Make sure you have as many references as possible to answer questions that you have as you move forward with your ERP solution. An ERP Help Desk can provide you with these resources-tools. They are made up of Internal ERP Solution Library, Internal ERP Subject Matter Experts-SME's, and External

Resource; the ERP solution provider and reseller. All are of the utmost importance and they will ensure that you have the necessary resources to eliminate barriers that you will encounter utilizing the ERP solution.

In closing, this guide is the first step in understanding the importance of the preparation, selection and the implementation of an ERP system. It is a team effort and requires everyone within the organization to be involved and part of the successful implementation process. Whenever there are doubts, the issues presented should be challenged and resolved. The involvement of your employees is critical and they should be supported so that they can be an integral part of the project. Do not skimp on education and training either! A well versed workforce is a key element in a successful ERP project. Last, data is king! Do not assume that the data you have is clean and usable! Migrate small pieces of your data from your legacy system and see if it works. Why bring over all of the data if the data has issues? Manage this in small segments and the process will be made much easier while reducing the time needed to migrate the data into the new ERP solution.

The authors wish you the best of success and we know that what was spoken about in this guide will help you navigate your ERP preparation, selection and implementation of your new ERP solution.

Please go to to the IMAE website, and use the “Ask theAuthors” service if you have questions and or want to let us know how the project went during the ERP implementation process.

www.instituteforimae.com

NOTES:

REFERENCES

O'Sullivan, J.A. (2014) Validating Academic Training versus Organizational Training Using ERP . ISBN 978-3-659-59768-8. Lambert Academic Publishing, Saarbrücken, Germany.

O'Sullivan, J.A., Caiola, G. (2008) Enterprise Resource Planning a Transitional Approach from the Classroom to the Business World. McGraw Hill Publishing, New York.

Ptak, C.A., Schragenheim, E. (2004) ERP : Tools, Techniques, and Applications for Integrating the Supply Chain. Boca Raton: St. Lucie Press.

Cronan, T.P., Douglas, D.E., Alnuaimi, O., Schmidt, P.J. (2009) 2009 ITRI, Working Paper Series, pp. 122-1008, pp. 123-1008.

Antonucci, L., zur Muehlen, M. (2009) Developing an International Business to Business Process Curriculum: Extending the Classroom Walls with ERP Software.

Flinchbaugh, J., Carlino, A. (2006) The Hitchhiker's Guide to the Lean Lesson Learned from the Road. pp. 55. Society of Manufacturing Engineers.

www.BusinessDictionary.com

WWW.THE9000STORE.COM

http://it.toolbox.com/blogs/inside-erp/is-your-erp-project-headed-for-failure-63204?reftrk=no&trdref=4232425265674d61696c696e673a496e736964652d455250&mailing_id=1059094, accessed 10/23/2014.

GLOSSARY

A

APICS-Founded in 1957 as the American Production and Inventory Control Society, APICS, The Association for Operations Management builds operations management excellence in individuals and enterprises through superior education and training, internationally recognized certifications, comprehensive resources, and a worldwide network of accomplished industry professionals.

B

Back Flushing (back flush costing)-The application of costs based on the output of a process. Back flush costing is usually associated with repetitive manufacturing environments.

Back Scheduling-A technique for calculating operation start dates and due dates. The schedule is computed starting with the due date for the order and working backward to determine the required start date and/or due dates for each operation. Syn: backward scheduling. Ant: forward scheduling.

Best in Class: The highest current performance level in an industry, used as a standard or benchmark to be equaled or exceeded. Also called, Best of Breed.

Bill of Material: (1) A listing of all sub-assemblies. Intermediates, parts, and raw materials that go into a parent assembly showing quantity of each required to make an assembly. It is used in conjunction with the master production schedule to determine the items for which purchase requisitions and production orders must be released. A variety of display formats exit for bills of material, including the single level bill of material, transient bill of material, matrix bill of material. (2) A list of all the materials needed to make one piece parts/components for its customer. The bill of material may also be called the formula, recipe or ingredients list in certain process industries.

Booking: Customer's concluded arrangement with a goods or services supplier (carrier, hotel, restaurant, publisher, etc.) representing a completed sale. Also called reservation or transaction.

Brainstorm(ing): A technique that teams use to generate ideas on a particular subject. Each person on the team is asked to think creatively and write down as many ideas as possible. The ideas are not discussed or reviewed until after the brainstorming session.

Buyer/Planner: A buyer who also does material planning. The term should not be confused with planner/buyer, which is a synonym for supplier scheduler.

C

Capacity Planning-The function of establishing, measuring, and adjusting limits or levels of capacity. The term capacity requirements planning in this context refers to the process of determining in detail the amount of labor and machine resources required to accomplish the tasks of production. Open shop orders and planned orders in the MRP system are input to CRP, which through the use of parts routings and time standards translates these orders into hours of work by work center by time period. Even though rough-cut capacity planning may indicate that sufficient capacity exists to execute the MPS, CRP may show that capacity is insufficient during specific time periods.

Capital Purchases-Capital purchases refer to the purchase of capital equipment, which can be a Lathe for the shop or a computer for the planning department.

Champion: Person who voluntarily takes extraordinary interest in the adoption, implementation, and success of a cause, policy, program, project, or product. He or she will typically try to force the idea through entrenched internal resistance to change, and will evangelize it throughout the organization. Also called change advocate, change agent, or idea champion.

Change Agent: Person who voluntarily takes extraordinary interest in the adoption, implementation, and success of a cause, policy, program, project, or product. He or she will typically try to force the idea through entrenched internal resistance to change, and will evangelize it throughout the organization. Also called change advocate, or idea champion.

Change Management: The business process that co• ordinates and monitors all changes to the business processes and applications operated by the business as well as to their internal equipment, resources, operating systems, and procedures. The change management discipline is carried out in a way that minimizes the risk of problems that will affect the operating environment and service delivery to the users.

Charter: A team charter is a written document that defines the team's mission, scope of operation, objectives, time frame, and consequences. Charters can be developed by top management and presented to teams, or teams can create their own charters and present them to top management. Either way, the top management's endorsement of a team's charter is a critical factor in giving the team the direction and protection it needs to succeed. (www.highperformanceteams.org)

Closed Loop System-MRP-A system built around material requirements planning that includes the additional planning processes of production planning (sales and operations planning), master production scheduling, and capacity requirements planning. Once this planning phase is complete and the plans have been accepted as realistic and attainable, the execution processes come into play. These processes include the manufacturing control processes of input-output (capacity) measurement, detailed scheduling and dispatching, as well as anticipated delay reports from both the plant and sup• pliers, supplier scheduling, and so on. The term closed loop implies not only that each of these processes is included in the overall system, but also that feedback is provided by the execution processes so that the planning can be kept valid at all times.

Configuration control - The function of ensuring that the product being built and shipped corresponds to the product that was designed and ordered. This means the correct features, customer options, and engineering changes have been incorporated and documented.

Cost of Goods Sold-COGS: refer to the carrying value of goods sold during a particular period. Costs are associated with particular goods using one of several formulas, including specific identification, first-in first-out (FIFO), or average cost. Costs include all costs of purchase, costs of conversion and other costs incurred in bringing the inventories to their present location and condition. Costs of goods made by the business include material, labor, and allocated overhead. The costs of those goods not yet sold are deferred as costs of inventory until the inventory is sold or written down in value. (Wikipedia)

Critical Data Elements-These are data inputs to the system, that without very high levels of accuracy, the output will be severely compromised. There is little or no room for error or inaccuracies here.

Continuous Improvement-The act of making incremental, regular improvements and upgrades to a process, or product, in the search of excellence. (APICS dictionary, 11th edition)

Conference Room Pilot- The conference room pilot utilizes the test data base created by your IT department and your steering committee to test system accuracy in all aspects of the operation and must be fully flushed out and accepted before going live.

Corrective Action: Identification and elimination of the causes of a problem, thus preventing their recurrence.

Cost of Quality-The cost associated with providing poor quality products or services. There are four categories of costs: (1) internal failure costs (costs associated with defects found before the customer receives the product or service); (2) external failure costs (costs associated with defects found after the customer receives the product or service); (3) appraisal costs (costs incurred to determine the degree of conformance to quality requirements); and (4) prevention costs (costs incurred to keep failure and appraisal costs to a mini• mum).

Critical Path-The Critical Path, the longest necessary path through a network of activities when respecting their interdependencies, which may be identified with the Program Evaluation and Review Technique and the Critical Path Method. (Wikipedia)

Cumulative Lead Time-The longest planned length of time to accomplish the activity in question. It is found by reviewing the lead time for each bill of material path below the item; whichever path adds up to the greatest number defines cumulative lead time. Syn: aggregate lead time, combined lead time, composite lead time, critical path lead time, stacked lead time. See: planning horizon, planning time fence.

Cumulative Manufacturing Lead Time-The cumulative planned lead time when all purchased items are assumed to be in stock. Syn: composite manufacturing lead time.**Current State:** Helps visualize the current production or administrative processes and identify sources of waste.

D

Dash Boards-Digital: An executive information system available for computers that offers a set of customizable windows and data portals. Information helpful to business managers available on a digital dashboard might include stock prices, weather data, a calendar management function, e-mail access, news feeds and various types of company specific information.

Data Migration -Data migration refers to the mapping of data from your current system to the new system and ensuring all data points are reflected correctly in it new field.

Data Mapping: is the process of matching data sets in your current software to the equivalent data sets in the new software. (From guide rough draft)

Direct Labor-Labor that is specifically applied to the good being manufactured or used in the performance of the service.

E

Engineering Change Notice-ECN: 1.Alteration made to a bill of materials by (1) adding or deleting an item (assembly, sub-assembly, component, or material), (2) substituting an item with another, or (3) changing the usage of an item. 2. Alteration made in the functional, maintenance, performance, or physical characteristics of a system. Also called task change.

Engineering Change Request form (ECR) - engineering change-A revision to a drawing or design released by engineering to modify or correct a part. The request for the change can be from a customer or from production, quality control, another department, or a supplier. Syn: engineering change notice, engineering change order. Engineering change notice-Syn: engineering change order-Syn: engineering change. Engineering characteristics-The technical features designed into a product

External Costs-Are those that affect processes that are outside the four walls of your organization. These are supply chain, logistics, replenishment schedules, lead time, and others. If there are internal and external errors, this greatly compounds the ability of the organization of delivering a quality product and or service on time and within the budget allotted.

F

Final Assemblies-The highest level assembled product, as it is shipped to customers.

Five Whys: Is a technique to try to ensure that the root causes of problems-waste are sought out. (The Lean Toolbox the Essential Guide to Lean Transformation)

Flowcharting: A flowchart is a type of diagram that represents an algorithm, workflow or process, showing the steps as boxes of various kinds, and their order by connecting them with arrows. This diagrammatic representation illustrates a solution model to a given problem. Flowcharts are used in analyzing, designing, documenting or managing a process or program in various fields.

Forecast Accuracy: A measurement of forecast usefulness, often defined as the average difference between the forecast values to the actual value.

Forecasted Demand - forecast error-The difference between actual demand and forecast demand, stated as an absolute value or as a percentage. See: average forecast error, forecast accuracy, mean absolute deviation, tracking signal.

Forward Scheduling or Finite Scheduling-A scheduling technique where the scheduler proceeds from a known start date and computes the completion date for an order, usually proceeding from the first operation to the last. Dates generated by this technique are generally the earliest start dates for operations. See: forward pass. Ant: back scheduling.

Future State: Helps visualize the future production or administrative processes and how the elimination of waste has improved the throughput of both the service and production processes.

G

Gap: When the actual performance level is not equal to the expected performance level.

Gap Analysis: A technique that businesses use to determine what steps need to be taken in order to move from its current state to its desired, future state. Also called need-gap analysis, needs analysis, and needs assessment. Gap analysis consists of (1) listing of characteristic factors (such as attributes, competencies, performance levels) of the present situation ("what is"), (2) listing factors needed to achieve future objectives ("what should be"), and then (3) highlighting the gaps that exist and need to be filled. Gap analysis forces a company to reflect on who it is and ask who they want to be in the future.

GIGO-Garbage In Garbage Out. This refers to the fact that as long as your data inputs are inaccurate or incorrect the data output will also be inaccurate or incorrect.

Goal: is a desired result a person or a system envisions plans and commits to achieve a personal or organizational desired end-point in some sort of assumed development. Many people endeavor to reach goals within a finite time by setting deadlines.

Going 'Live'-The term Going Live refers to the date and time you will be switching over to the new ERP/MRP system.

Going Parallel- Going parallel or running parallel refers to the running of your new ERP system at the same time you are still running the old system. This is very difficult to do and manage as you are dealing with two systems that need to be maintained.

I

Indentured Levels: A hierarchical BOM is a form of product breakdown structure because it captures product component relationships as well as identity and quantity. (www.action-engineering.com)

Indirect Labor- indirect labor-Work required to support production in general without being related to a specific product

Infinite Loading-Calculation of the capacity required at work centers in the time periods required regardless of the capacity available to perform this work. Syn: infinite scheduling.

Internal Customer-Your internal customers refer to the department you are serving whether it be direct labor or indirect labor oriented. For example the Machine shop provide parts to the Production floor. In this case the Production floor is the customer of the Machine shop. Another example would be purchasing buying part for the Engineering department. The engineering department is the customer of purchasing.

Interventions: A method of introducing planned change.*

Inventory Carrying Costs- The cost of holding inventory, usually defined as a percentage of the dollar value of inventory per unit of time (generally one year).

ISO-International Organization for Standardization

ISO Certified Company-A company that conforms to the guidelines set forth by the stringent quality requirements of the International Standardization Organization

Inventory Carrying Costs: The cost of holding inventory, usually defined as a percentage of the dollar value of inventory per unit of time (generally one year).

K

KPI-Key Performance Indicators: A financial or nonfinancial measure, either tactical or strategic, that is linked to specific strategic goals and objectives.

Kaizen-The Japanese term for "Improvement; "continuing improvement involving everyone including managers and workers.

Kaizen Blitz-A rapid improvement of a limited process area, for example, a production cell. Part of the improvement team consists of workers in that area. The objectives are to use innovative thinking to eliminate non-value-added work and to immediately implement the changes within a week or less. Ownership of the improvement by the area work team and the development of the team's problem-solving skills are additional benefit.

Kaizan Event-A time-boxed set of activities carried out by the cell team during the week of cell implementation. The kaizen event is an implementation arm of a lean manufacturing program.

L

Lead Times: (1) A span of time required to perform a process (or series of operations). In logistics, context, the time between recognition of the need for an order and a receipt of goods. Individual components of lead time can include preparation time, queue time, processing time, move or transportation time and receiving and inspection time.

Legacy Data-Legacy data is data that exists in your current ERP/MRP system.

Legacy System-A Legacy system is the current system you are presently using to run your business.

Lean Manufacturing-A philosophy of production that emphasizes the minimization of the amount of all the resources (including time) used in the various activities of the enterprise. It involves identifying and eliminating non-value-adding activities in design, production, supply chain management, and dealing with customers. Lean producers employ teams of multi-skilled workers all levels of the organization and use highly flexible, increasingly automated machines to produce volumes of products in potentially enormous variety. It contains a set of principles and practices to reduce cost through the relentless removal of waste and through the simplification of all manufacturing and support processes.

Linearity-Production at a constant quantity and use of resources at a level rate, typically measured daily, or more frequently.

M

Manufacturing Resource Planning (MRP II): A method for the effective planning of all resources of a manufacturing company. Ideally, it addresses operational planning in units financial planning dollars, and has a simulation capability to answer what-if questions. It is made up of processes, each linked together: business planning capacity requirements planning, and the execution support systems for capacity and material. Output from these systems is integrated with financial reports such as the business plan, purchase commitment support, shipping budget, and inventory projections in dollars. Manufacturing resource planning is a direct outgrowth and extension of closed-loop MRP.

MPS Master Production Schedule (MPS)-The master production schedule is a line on the master schedule grid that reflects the anticipated build schedule for those items assigned to the master scheduler. The master scheduler maintains this schedule, and in tum, it becomes a set of planning numbers that drives material requirements planning. It represents what the company plans to produce expressed in specific configurations, quantities, and dates. The master production schedule is not a sales item forecast that represents a statement of demand. The master production schedule must take into account the forecast, the production plan, and other important considerations such as backlog, availability of material availability of capacity, and management policies and goals. See: master schedule.

Material Requirements Planning (MRP)-A set of techniques that uses bill of material data, inventory data, and the master production schedule to calculate requirements for materials. It makes recommendations to release replenishment orders for material. Further, because it is time-phased, it makes recommendations to reschedule open orders when due dates and need dates are not in phase. Time-phased MRP begins with the items listed on the MPS and determines (1) the quantity of all components and materials required to fabricate those items and (2) the date that the components and material are required. Time-phased MRP is accomplished by exploding the bill of material, adjusting for inventory quantities on hand or on order, and offsetting the net requirements by the appropriate lead times.

Material Variances-Material variances occur when there is a difference between how much you actually used to produce and item and how much you were should have used based on the bill of material and or the routing.

Matrix: A mathematical array having one, two, and sometimes more dimensions into which collections of data may be stored and processed.

Metric- Performance Measure: In a performance measurement system that actual value measured for criterion.

Mission: is a statement of the purpose of a company, organization or person its reason for existing. The mission statement should guide the actions of the organization, spell out its overall goal, provide a path, and guide decision making. It provides "the framework or context within which the company's strategies are formulated." It's like a goal for what the company wants to do for the world.

Muda: Is the Japanese term for waste. (The Lean Toolbox the Essential Guide to Lean Transformation)

N

Non-Value Added-The term Non-Value added refers to any action, process, or procedure that doesn't add to the inherent value of the part or process. Non-Value added activities should be eliminated wherever possible.

P

Perpetual Inventory-Inventory recordkeeping system is where each transaction in and out is recorded and a new balance is computed.

Pilot (Test): (1) in computer systems a test before final acceptance of a new business system using a subset of data with engineered cases and documented results. (2) Generally, production of a quantity to verify manufacturability, customer acceptance, or other management requirements before implementation of ongoing production.

Process: A series of actions, changes, or functions bringing about a result. (Six Sigma Development Solutions)

Process Improvement: is an aspect of Organizational Development (OD) in which a series of actions are taken by a process owner to identify, analyze and improve existing business processes within an organization to meet new goals and objectives such as increasing profits and performance, reducing costs and accelerating schedules. These actions often follow a specific methodology or strategy to increase the likelihood of successful results. Process improvement may include the restructuring of company training programs to increase their effectiveness. Process improvement is also a method to introduce process changes to improve the quality of a product or service. (Wikipedia)

Process Owner: The process owner is responsible for designing the processes necessary to achieve the objectives of the business plans that are created by the Business Leaders. The process owner is responsible for the creation, update and approval of documents (procedures, work instructions/protocols) to support the process. Many process owners are supported by a process improvement team. The process owner uses this team as a mechanism to help create a high performance process. The process owner is the only person who has authority to make changes in the process and manages the entire process improvement cycle to ensure performance effectiveness. This person is the contact person for all information related to the process. This person is accountable for the effectiveness of the process. (Wikipedia)

Product Families: Group of products derived from a common product platform. These goods or services use similar or same production processes, have similar physical characteristics, and may share customer segments, distribution channels, pricing methods, promotional campaigns, and other elements of the marketing mix. Products comprising a family are usually priced and discounted as a package. Several product families make up a product portfolio. Also called product group or product line.

Product Structure -A computer record defining the relationship of one component to its immediate parent and containing fields for quantity required, engineering affectivity, scrap factor, application selection switches, and so forth

Productivity: (1) An overall measure of the ability to produce a good or service. It is the actual output of production compared to the actual input of resources. Productivity is a relative measure across time against common entities (labor, capital, etc.). In production literature, attempts have been made to define total productivity where the effects of labor and capital are combined and divided into output. One example is a ratio that is calculated by adding the dollar value of labor, capital equipment, energy and material and so forth and dividing it into the dollar value of output in a given time period. This is one measure of total factor productivity. (2) In economics, the ratio of output in terms of dollars of sales, to an input, such as direct labor, in terms of the total wages. This is called the single factor productivity or partial factor productivity.

Profit Margin: Ratio of profit after taxes to cost-of-sales, often expressed as a percentage. It is one of the measures of the profitability of a firm, and an indicator of its cost structure. Formula: After-tax profit x 100 ÷ cost of sales.

Project Management Software: has the capacity to help plan, organize, and manage resource pools and develop resource estimates. Depending on the sophistication of the software, it can manage estimation and planning, scheduling, cost control and budget management, resource allocation, collaboration, communication, decision making, quality management and documentation or administration systems. Today, numerous PC & browser based project management software exist and they are finding their way into almost every type of business. (Wikipedia)

Project Plan: according to the Project Management Body of Knowledge is: "...a formal, approved document used to guide both project execution and project control. The primary uses of the project plan are to document planning assumptions and decisions, facilitate communication among stakeholders, and document approved scope, cost, and schedule baselines. A project plan may be summarized or detailed. (Wikipedia)

Purchase Price Variance-PPV: The difference for an item or service between the amount per unit actually paid and the budgeted amount per unit multiplied by the number of units bought. The purchase price variance of input materials can significantly affect production costs for a manufacturing business that does not have control over the cost of its production inputs.

Q

Quality Management System (QMS)-Is a collection of business processes focused on achieving your quality policy and quality objectives — i.e. what your customer wants and needs.[1] It is expressed as the organizational structure, policies, procedures, processes and resources needed to implement quality management. Early systems emphasized predictable outcomes of an industrial product production line, using simple statistics and random sampling. By the 20th century, labor inputs were typically the most costly inputs in most industrialized societies, so focus shifted to team cooperation and dynamics, especially the early signaling of problems via a continuous cycle. In the 21st century, QMS has tended to converge with sustainability and transparency initiatives, as both investor and customer satisfaction and perceived quality is increasingly tied to these factors. Of all QMS regimes, the ISO 9000 family of standards is probably the most widely implemented worldwide – the ISO 19011 audit regime applies to both, and deals with quality and sustainability and their integration. (Source Wikipedia)

R

Relational Database- A software program that allows users to obtain information drawn from two, or more, databases that are made up of two-dimensional arrays of data.

Return on Investment-ROI: The earning power of assets measured as the ratio of the net income (profit less depreciation) to the average capital employed (or equity capital) in a company or project. Expressed usually as a percentage, return on investment is a measure of profitability that indicates whether or not a company is using its resources in an efficient manner. For example, if the long-term return on investment of a company is lower than its cost-of-capital, then the company will be better off by liquidating its assets and depositing the proceeds in a bank. Also called rate of return, or yield.

Robert's Rules of Order: Provides common rules and procedures for deliberation and debate in order to place the whole membership on the same footing and speaking the same language. The conduct of ALL business is controlled by the general will of the whole membership - the right of the deliberate majority to decide. Complementary is the right of at least a strong minority to require the majority to be deliberate - to act according to its considered judgment AFTER a full and fair "working through" of the issues involved. Robert's Rules provides for constructive and democratic meetings, to help, not hinder, the business of the assembly. Under no circumstances should "undue strictness" be allowed to intimidate members or limit full participation. http://www.robertsrules.org

Rough Cut Capacity Planning (RCCP) -Rough cut capacity Planning tries to determine the fit of load versus available capacity on a more global scale. It looks at the load represented by the forecast in total against the capacity of the entire business and does not take into account the possibility that there may be individual machine or labor constraints specific to an individual area.

S

Sales and Operation Planning, S & OP: A process to develop tactical plans that provide management the ability to strategically direct its businesses to achieve competitive advantage on a continuous basis by integrating customer focused marketing plans for new and existing products with the management of the supply chain. The process brings together all the plans for the business (sales, marketing, and development, manufacturing, sourcing, and financial) into one integrated set of plans. It is performed once a month and is reviewed by management an aggregate (product family) level. The process must reconcile all supply, demand, and new product plans for both the detail and aggregate levels and tie the business plan. It is the definitive statement of the company's plans for the near to intermediate term, covering a horizon sufficient to plan for. (APICS Dictionary 13th edition)

Sales Forecast Accuracy- forecast accuracy-A measurement of forecast usefulness, often defined as the average difference between the forecast values to the actual value. Syn: sales forecast. .See: forecast error.

Scope-In project management, subdividing a project into smaller components to facilitate management.

Scope creep (also called requirement creep and feature creep) in project management refers to uncontrolled changes or continuous growth in a project's scope. This can occur when the scope of a project is not properly defined, documented, or controlled. It is generally considered harmful.

Scribe: A person who is part of a project team and takes minutes of the meeting proceedings defined by the meeting agenda. http://www.robertsrules.org/

SME-Subject Matter Expert

Soft Skills: Communicating, conflict management, human relations, making presentations, negotiating, team building, and other such ability, defined in terms of expected outcomes and not as a specific method or technique such as statistical analysis.

SOP- Standard Operating Procedure.

Sponsor: Individual or entity who organizes and is committed to the development of a product, program, or project.

Stakeholders-People with a vested interest in a company, including managers, employees, stockholders, customers, suppliers, and others.

Standard Costs -The target costs of an operation, process, or product including direct material, direct labor, and overhead charges

Standard Time- The length of time that should be required to (1) set up a given machine or operation and (2) run one batch or one or more parts, assemblies, or end products through that operation. This time is used in determining machine requirements and labor requirements. Standard time assumes an average worker following prescribed methods and allows time for personal rest to overcome fatigue and unavoidable delays. It is also frequently used as a basis for incentive pay systems and as a basis of allocating overhead in cost accounting systems. Syn: Standard Hours.

Standard Work-Standard work is work that is performed within the guidelines of written procedures and/or methods.

Steering Committee: It is a body within an enterprise that supports the steering of its actions. Its main concern is making strategic decisions concerning future realization of the enterprise's investment projects. It makes decisions about which of the presented projects will be realized, and which will not. It is responsible for the management and monitoring of a long-term project, which means that it controls the realization of the project at the strategic level, verifies the project's coherence with established aims, and keeps established frames such as range, costs and deadlines. Should any changes in the project happen, they must be first presented and accepted by a member of the steering committee. Only in the case of acceptance are the changes introduced to the project. The steering committee assesses and accepts the changes by means of consensus. It deals also with coordination and coherence with other realized projects. The steering committee creates working groups and chooses experts, with whom the enterprise will work to realize the project. The manager of a project is accountable for a project in front of the committee. (Wikipedia)

System Variables-System variables are switch in the ERP software that determines how data will be looked, processed and outputted. It is critical that all involved be aware of how these switches are set. They can significantly affect ERP/MRP output.

Supply Chain: is a system of organizations, people, activities, information, and resources involved in moving a product or service from supplier to customer. Supply chain activities transform natural resources, raw materials, and components into a finished product that is delivered to the end customer. In sophisticated supply chain systems, used products may re-enter the supply chain at any point where residual value is recyclable. Supply chains link value-chains. (Wikipedia)

SWOT Analysis- An analysis of the strengths, weaknesses, opportunities, and threats of and to an organization. SWOT analysis is useful in developing strategy.

T

Timeline: is a way of displaying a list of events in chronological order, sometimes described as a project artifact. It is typically a graphic design showing a long bar labeled with dates alongside itself and usually events labeled on points where they would have happened. (Wikipedia)

Time phasing-The technique of expressing future demand, supply, and inventories by time period. Time phasing is one of the key elements of material requirements planning.

Time phased receipt of material- time-phased order point (TPOP)-MRP-like time planning logic for independent demand items, where gross requirements come from a forecast, not via explosion. This technique can be used to plan distribution center inventories as well as to plan for service (repair) parts, because MRP logic can readily handle items with dependent demand, independent demand, or a combination of both. Time-phased order point is an approach that uses time periods, thus allowing for lumpy withdrawals instead of average demand. When used in distribution environments, the planned order releases are input to the master schedule dependent demands. See: fixed reorder quantity inventory model.

V

Value added-1) In accounting, the addition of direct labor, direct material, and allocated overhead assigned at an operation. It is the cost roll-up as a part goes through a manufacturing process to finished inventory. 2) In current manufacturing terms, the actual increase of utility from the viewpoint of the customer as a part is transformed from raw material to finished inventory. It is the contribution made by an operation or a plant to the final usefulness and value of a product, as seen by the customer. The objective is to eliminate all non-value-added activities in producing and providing a good or service.

Value-added Network (VAN)-A network, often supporting EDI, providing services additional to those provided by common carriers.

Vision: The shared perception of the organization's future-what organization will achieve and a supporting philosophy. This shared vision must be supported by strategic objectives, strategies, and an action plan to move it in the desired direction.

W

Work Instruction-A work instruction is similar to a procedure except that it is far more detailed and has within the step by step, down to the keystroke, instruction on how to perform an individual task. It can be used as a training tool for new employees.

Work in Process (WIP)-Work in process is exactly what it sounds like. It is work that has been released to the shop floor to be manufactured and is considered to be in process until it is complete.

Z

Zero Level: Is the first level of a BOM and represents the finish goods item. (www.it.toolbox.com)

FIGURES AND TABLES

The Convention for the Figures and Tables is as follows:

CPSIA information can be obtained at www.ICGtesting.com
Printed in the USA
BVIW12n1409040917
493897BV00010B/240